喚醒你的英文語感 ！

Get a Feel for English!

商務英文
EMail
速成語庫書

作者 ◎ 商英教父 Quentin Brand

作者序

　　本人拙作《愈忙愈要學英文 Email》出版至今已逾十年，而從那時起，台灣的經營環境也產生了極大的變化！我在 2004 年寫書的時候，電子郵件還是新的通訊系統，很多人搞不清楚要怎麼用。一種新的通訊就需要一種新的學習方式。如今十年過去了，人人都在寫電子郵件，寫起來也覺得更加自在與自信。

　　《愈忙愈要學英文 Email》是台灣第一本英文電子郵件寫作書；2011年我們又出版了一本方便攜帶的口袋書《愈忙愈要學英文字串 Email篇》，其中收錄了許多實用的 Email 字串。而我衷心認爲，這兩本書對於協助讀者寫電子郵件時建立信心扮演了小小的角色。我要謝謝所有曾經購買過去那兩本書和持續購買本書的人。我尤其要感謝我當時和現在的學生，我從他們身上學到了很多，他們也不斷在啓發和教導我。

　　本次新書匯集了兩本 Email 寫作書，兼具學習與工具的需求，並搭配貝塔官網線上支援，讀者能夠輕鬆地利用桌上型或平板電腦取得加值資源。此外，還新增了練習題，因爲各位永遠都嫌練習不夠，對吧？

　　誠摯希望新一代的讀者也會覺得本書很有用。

Quentin Brand

It's now been ten years since the publication of 愈忙愈要學英文 Email and since that time so much has changed in the business environment in Taiwan! When I wrote the book in 2004, email was a new system of communication, and many people were confused about how to deal with it. A new kind of comunication needed a new kind of learning approach. Now, ten years later, everyone is writing emails and feels more comfortable and confident doing so.

This book was the first email writing in English book to be published in Taiwan. In 2011 we published a convenient pocket book of useful email set-phrases 愈忙愈要學英文字串 Email 篇. I'd like to think that these two books played a small part in helping people to build their confidence with writing emails. I want to thank all the people who bought copies, and who continue to buy copies. In particular I want to thank my students –then and now– from whom I have learnt so much, and who continue to inspire me and teach me.

This new edition combines both email books and uses e-learning platform that you can easily access from computer or tablet. It also contains updated practice materials, because you can never have enough practice, right?!

I sincerely hope a new generation of readers will also find the new book useful.

Quentin Brand

C @ N T E N T S

Section 1　學習篇：基礎字串概念

✉ Unit 1・商務 Email 寫作的基本要領

✉ Unit 2・開場白 Set-phrases

✉ Unit 3・表達請求的 Email

Section 2 工具篇：好用句型及例句

✉ Unit 8 · 發信的開場白

✉ Unit 9 • 回信的開場白

✉ Unit 10 • 表達請求

✉ Unit 11 • 安排會面

✉ Unit 12 • 急件

✉️ Unit 13 • 描述和解決問題

✉️ Unit 14 • 客戶管理與維繫

✉️ Unit 15 • 關於附件

✉️ Unit 16 • 結語

前 言

現今電子郵件已成為全球商業界最普及的溝通方式。一天當中，一家國際公司的主管會收發十幾二十封的電子郵件，不管是寄給國外子公司、客戶、消費者、地區總公司，甚至同公司同部門的同事，大多數人都以英文書寫電子郵件。由此可見，公司上上下下從初階的實習生到高級主管，只要是商界人士，英文電子郵件寫作已成為一項不可或缺的技能。然而許多人卻苦於寫不出清楚明確的電子郵件，而在工作中倍受挫折。據估計，為了語意不清的英文電子郵件，一人每天至多會浪費一個半小時的上班時間，彼此反反覆覆地解釋和釐清郵件的原義，才能處理好事情。因此本書的兩大目標，即為教你英文電子郵件簡單又實際的寫作方法，以節省時間，同時加強你的英文能力。

現在請花點時間回答以下問題。請即刻作答，答畢再往下看。

Task 1

請回答以下問題，並寫下答案。

1. 為何購買本書？
2. 希望從本書學到什麼？
3. 寫英文電子郵件時，遇到哪些困難？
4. 電子郵件的風格為何？

請參考下列可能答案，勾出最接近你的答案者：

1. 為何購買本書？

　□ 我買這本書是為了學英文，用於專業。

　□ 我是個大忙人，不想浪費時間學工作上用不到的東西，更不想做與工作無關的英文練習。

□ 因為我在這本書的封面上看到「英文 Email」一詞。市面上的書大多不是關於一般商業寫作就是關於商業信件，我從來沒有寫商業書信的需要，這類書籍根本派不上用場！更何況，電子郵件和商業企劃案全然是兩回事，兩者的寫作技巧不同。

□ 我要一本教我寫電子郵件的書。我希望這本書以練習引導學習，提供簡單易懂的參考資料，像英文字典一樣，可以放在電腦旁隨時翻閱。

□ 我要一本了解我需求的書！

2. 希望從本書學到什麼？

□ 我想學到工作上需要用到的常見詞彙和文法。

□ 電子郵件有特定格式嗎？如果有，我想學。

□ 我希望這本書像英文家教一樣，幫我指出錯誤，予以糾正。

□ 我想學好英文。我的客戶遍及英國、美國，還有歐洲、印度，甚至東南亞，我希望他們都能看懂我寫的電子郵件。

□ 我同一件事只有一種寫法，覺得自己寫的電子郵件肯定枯燥乏味。我想學會活用文字。

□ 我不大會唸英文，更討厭文法。文法在我看來無聊透頂，比起站在英文母語人士面前做簡報還可怕！可是偏偏文法很重要，所以希望能找到加強英文電子郵件寫作能力，卻不必苦讀文法的方法。

□ 我想找到自學英文的方法。我在英語的環境中工作，但自知沒有善加利用這個優勢，加強專業英語能力。希望這本書可以教我如何靠自己加強英文。

3. 寫英文電子郵件時，遇到哪些困難？

□ 寫電子郵件時，我常常不知該如何起頭。有時我得坐在書桌前對著電腦乾瞪眼，想半天才想出怎麼起頭。

□ 我不確定我寫的電子郵件是否言辭達意，文字準不準確，結構對不對。對於對方是否看懂我的意思，我沒信心。

□ 高中時學到的文法規則一籮筐，但有一大半已經還給老師，真希望當時有專心上課！

□ 有時我絞盡腦汁，仍然不知道該如何措辭以適當地表達意思。

☐ 我寫一封電子郵件得耗上大把時間。我真的很想提高效率，節省時間做其他事。

☐ 文法、拼字，我通通不懂！

4. 電子郵件的風格為何？

☐ 這問題很難回答。大學時學過商業書信寫作，所以知道商業書信的風格為何，但是電子郵件我就不曉得了。

☐ 有時會擔心自己寫的電子郵件風格不對，客戶一定瞧不起我。但我認為重點是事情完成了就好，這樣的觀念對嗎？

☐ 話怎麼講，電子郵件就怎麼寫，不是嗎？

你可能對上面的部分答案，或所有答案有同感，也可能有其他想法，不過請先容我自我介紹。我是 Quentin Brand，過去二十五年來在全球各地從事英語教學，其中大半時光是待在台灣教書，我的教學對象便是像你這樣的商界專業人士。從大型國際企業的國外分公司經理，到有國外市場的小型國內公司的初階實習生，我的學生跨足商界各階層。每個學生均曾吐露如上所述的心聲，他們（包括你）共同的心願，就是找到既簡單又實際的方法學英文。

這次你可是找對人了！多年來我針對忙碌的商界人士，研究出一套以嶄新角度看語文的英文教學法，其核心概念稱作 Leximodel。現今全球一些屬一屬二的大公司均利用 Leximodel 幫助主管充分開發英文潛能。而本書的教學基礎，正是 Leximodel。

前言的教學目標是介紹和教你運用 Leximodel。我也會解釋本書用法，以及如何從本書獲得最佳學習效果。看完本章，你應達到的學習目標如下：

☐ 清楚了解 Leximodel 為何，以及用 Leximodel 學英文的好處。

☐ 了解 chunks、set-phrases 和 word partnerships 的差別。

☐ 閱讀書信時，能夠辨認文中的 chunks、set-phrases 和 word partnerships。

☐ 知道學習 set-phrases 時會遇到哪些困難，以及如何克服這些困難。

☐ 清楚了解本書中的不同要素，以及如何運用這些要素。

在繼續往下看之前，我要談談本書中 Task 對學習的重要性。相信你已經在前面的部分中注意到，我會請你暫停下來做 Task，有時還得將答案寫下來，在往後的章節中，我也會要求你先將 Task 做完再往下看。盼你能夠照做。

本書中的各章節都有許多 Task，這些 Task 皆經過精心設計，往往可刺激你下意識地消化新習得的語文。做 Task 時的思考過程遠比答對問題重要，**因此請務必循序漸進地做 Task，作答完畢之前切勿先看答案。**當然，為了節省時間，你大可不做 Task，一股作氣讀完整本書，但是事實上沒有動腦做書中的 Task，就達不到最佳學習效果，如此不啻是浪費時間罷了。請相信我的話，按部就班做 Task 準沒錯！

The Leximodel

可預測度

Leximodel 是從全新角度看語言的英語教學法，所根據的概念很簡單：

Language consists of words which appear with other words.
語言由字串構成。

這概念簡單易懂，看似連小學生都明白的道理。其實 **Leximodel 的基礎概念就是從字串的層面來看語言，而非以文法和單字**。為了讓你了解我的意思，我們來做一個 Task 吧！現在請做 Task 2，作答完畢再往下看。

Task 2

想一想，平常下列單字後面都會搭配什麼字？請寫在空格中。

listen	_____
depend	_____
English	_____
financial	_____

第一個單字旁你寫的是 to，第二個字旁寫的是 on，我猜得沒錯吧？我怎麼會知道？因為一種稱作「語料庫語言學」（corpus linguistics）的軟體程式和電腦技術做過語言分析之後，發現 listen 後面接 to 的機率非常高（98.9%），depend 後面接 on 的機率也相差不遠。換句話說，listen 和 depend 兩字後面接的字幾乎千篇一律，不會改變（listen 後接 to；depend 後接 on）。由於機率非常高，這兩組詞可視為 fixed（固定字串），也由於這兩組詞確實是固定的，當書寫 listen 和 depend 兩字時，後面沒有接 to 和 on，就可以說是寫錯了。

接下來的兩個字——English 和 financial ——後面該接什麼字比較難預測，所以我猜不出你在那兩個單字旁寫了什麼字。我可以在某個特定範圍內猜，你可能在 English 旁寫的是 class、book、teacher、email 或 grammar 等字；fnancial 旁寫的是 department、news、planning、product 或 stability 等字，卻無法像方才對前兩字那麼篤定了。原因何在？因為以統計預測 English 和 financial 後面接什麼字，準確率相對地低很多，很多字都有可能，而且每個字的機率相當。因此 English 和 financial 的字串可以說是不固定的，稱之為 fluid（流動字串）。由此推斷，語言不見得非以文法和單字來看不可，你大可將語言視作一個龐大的語料庫，裡面的字串有的是固定的，有的是流動的。根據可預測度，我們能看出字串的固定性和流動性，見圖示：

〈The Spectrum of Predictability 可預測度〉

字串的可預測度即為 Leximodel 的基礎，因此 Leximodel 的定義可以再追加一句話：

Language consists of words which appear with other words. These combinations of words can be placed along a spectrum of predictability, with fixed combinations at one end, and fluid combinations at the other.

語言由字串構成。這些字串可根據可預測度的程度區分，可預測度愈高的一端是固定字串，可預測度愈低的一端是流動字串。

◎ Chunks、set-phrases 和 word partnerships

你可能在心裡兀自納悶：我曉得 Leximodel 是什麼了，可是這對學英文有何幫助？我怎麼知道哪些是固定字串，哪些是流動字串？就算知道了，學英文會比較簡單嗎？別急，放輕鬆，從今天起英文會愈學愈上手！

字串（multi-word items，以下簡稱 MWI）可分成三大類：chunks、set-phrases 和 word partnerships。這些名詞沒有對等的中譯，因此請務必記得英文名稱。

現在我就開始詳細介紹各類字串吧。你很快就會發現各類字串都易懂而好用。首先來看看第一類 MWI——chunks。Chunks 的字串有固定也有流動元素。... listen to ... 即為一個很好的例子：listen 後面總是接 to，此乃其固定元素；但有時 ... listen ... 可以是 ... are listening ...、... listened ...、... have not been listening carefully enough ...，這些則是 listen 的流動元素。... give sth. to sb. ... 是另一個很好的例子：這裡的 give 後面得接某物（sth.），然後接某人（sb.）。因此 ... give sth. to sb. ... 在這裡是固定字串。不過 ... give sth. to sb. ... 這個 chunk 中，sth. 和 sb. 這兩個位置可以使用的詞範圍可就廣了，例如 give a raise to your staff（給員工加薪）和 give a presentation to your boss（向老闆做簡報）。看下面的圖你就懂了。

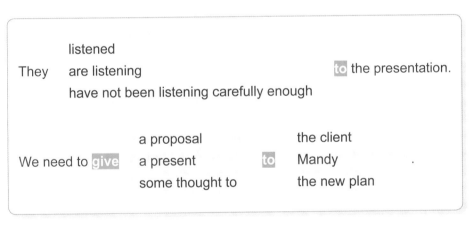

■ 部分為 fixed　　部分為 fluid

相信你能夠舉一反三，想出更多例子。當然，... give sth. to sb. ... 也可寫成 give sb. sth.，但 give sb. sth. 本身又是另一個 chunk 了，同樣是固定和流動的元素兼具。看得出來嗎？

Chunks 通常很短，為 meaning words（意義字，如 listen、depend）加上 function words（功能字，如 to、on）的組合。再做一個 Task 吧，看看是不是都懂了。注意，務必先做完 Task 再看答案，千萬不要作弊喔！

Task 3

請閱讀下面短文，找出所有的 chunks，並畫底線。

Everyone is familiar with the experience of knowing what a word means, but not knowing how to use it accurately in a sentence. This is because words are nearly always used as part of an MWI. There are three kinds of MWI. The first is called a chunk. A chunk is a combination of words which is more or less fixed. Every time a word in the chunk is used, it must be used with its partner(s). Chunks combine fixed and fluid elements of language. When you learn a new word, you should learn the chunk. There are thousands of chunks in English. One way you can help yourself to improve your English is by noticing and keeping a database of the chunks you find as you read. You should also try to memorize as many as possible.

翻譯

每個人都有這樣的經驗：知道一個字的意思，卻不知如何正確地用在句子中，這是因為每個字幾乎都必須當作 MWI 的一部分。MWI 可分為三類，第一類叫作 chunk。Chunk 幾乎是固定的字串，每當用到 chunk 的其中一字，該字的詞夥也得一併用上。Chunk 包含了語言中的固定元素和流動元素。學會一個新單字時，應該也連帶學會它的 chunk。英文中有成千上萬的 chunks。閱讀時留意並記下所有的 chunks，將之彙整成語庫，最好還要盡量背起來，不失為加強英文的好方法。

Task 3　參考答案

現在請以下列語庫核對答案，如果沒找到那麼多 chunks，可再看一次短文，看看是否能夠找到語庫中所有的 chunks。

• be familiar with n.p. ...	• every time + n. clause
• experience of Ving ...	• be used with n.p. ...
• how to V ...	• combine sth. and sth. ...
• be used as n.p. ...	• elements of n.p. ...
• part of n.p. ...	• thousands of n.p. ...
• there are ...	• in English ...
• kinds of n.p. ...	• help yourself to V ...
• the first ...	• keep a database of n.p. ...
• be called n.p. ...	• try to V ...
• a combination of n.p. ...	• as many as ...
• more or less ...	• as many as possible ...

💡 語庫小叮嚀

- 語庫中的 chunks，be 動詞都以原形 be 表示，而非 is 或 are。
- 記下 chunks 時，前後都加上 ...（刪節號）。
- 有些 chunks 後面接 V（go、write 等原形動詞）或 Ving（going、writing 等），有的則接 n.p.（noun phrase，名詞片語）或 n. clause（名詞子句）。我於「本書架構與使用說明」中會對此詳細解說。

　　好，接下來我要討論第二類 MWI：set-phrases。set-phrases 比 chunks 固定，通常字串較長，其中可能同時包含多個 chunks。**Chunks 大都是沒頭沒尾的片段文字組合。但是 set-phrases 通常包括句子的句首或句尾，甚至兩者兼具；換句話說，有時 set-phrases 會是一個完整的句子。**電子郵件中常可見到 set-phrases。請做下一個 Task。

📑 Task 4

下列語庫是電子郵件中常見的 set-phrases。請勾出你認識的 set-phrases。

- Thank you for sending me n.p. ...
- Apologies for the delay in getting back to you, but ...
- Thanks for your reply.
- Just to let you know that + n. clause ...
- Just to confirm that + n. clause ...
- Please confirm that + n. clause ...
- I look forward to hearing from you.
- If you have any questions, please do not hesitate to call either ... or ...
- If you have any questions about this, please do not hesitate to contact me.

💡 語庫小叮嚀

- 三類 MWI 中，固定性最高的就是 set-phrases，因此學習時務必鉅細靡遺地留意其中所有細節。稍後我會詳細解釋原因。

- 注意有些 set-phrases 以 n.p. 結尾，有的則以 n. clause 結尾。稍後我再對此詳述。

　　學會 set-phrases 的一大優點，就是在用法上絲毫不須費神操心文法問題，只要把本書的 set-phrases 當作固定字串背起來，原原本本照用即可。本書大部分的 Task 都和 set-phrases 有關，所以我會在下一節中對此作更詳細的解說。現在我們先來看看第三類 MWI：word partnerships。

　　三類 MWI 中，當屬 word partnerships 的流動性最高，其中含兩個以上的意義字（不同於 chunks 含意義字和功能字），並且通常是「動詞＋形容詞＋名詞」的組合。所有產業用的 chunks 和 set-phrases 都一樣，但是不同產業用的 word partnerships 就不同了。舉個例子，如果你在金融界中工作，用到的 word partnerships 就不同於資訊科技界人士。做完下面的 Task，你就會明白我的意思。

請看下列各組字串，請依據其 word partnerships 判斷各組所代表的產業，將答案寫在空格中。見範例①。

①

- government regulations
- drug trial
- patient response

- hospital budget
- key opinion leader
- patent law

產業名稱：_____醫藥界_____

②

- risk assessment
- non-performing loan
- credit rating

- share price index
- low inflation
- bond portfolio

產業名稱：_____

③

- bill of lading
- shipment details
- customs delay

- shipping date
- letter of credit
- customer service

產業名稱：_____

④

- latest technology
- user interface
- system problem

- repetitive strain injury
- input data
- installation wizard

產業名稱：_____

② 銀行和金融業　　　③ 外銷／進口業　　　④ 資訊科技業（IT）

　　如果你在上面其中一個產業中工作，一定早就認出其中一些 word partnerships 了吧？

　　現在 Leximodel 的定義應該要修正了：

Language consists of words which appear with other words. These combinations can be categorized as chunks, set-phrases and word partnerships and placed along a spectrum of predictability, with fixed combinations at one end, and fluid combinations at the other.

　　語言由字串構成，所有的字串可分成三大類──chunks、set-phrases 和 word partnerships，並且可依其可預測的程度區分，可預測度愈高的一端是固定字串，可預測度愈低的一端是流動字串。

　　新的 Leximodel 圖示如下：

〈The Spectrum of Predictability 可預測度〉

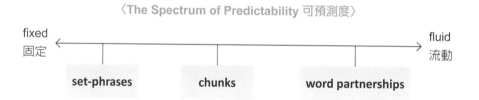

　　學英文致力於學好 chunks，文法自然會進步，因為大部分的文法錯誤都源自於 chunks 寫錯。學英文時專攻 set-phrases，英語功能會加強，因為 set-phrases 都是功能性字串。學英文時在 word partnerships 下功夫，字彙會增加。因此最後的 Leximodel 圖示如下：

〈The Spectrum of Predictability 可預測度〉

Leximodel 的優點與其對學英文的妙用，就在於無論說、寫英文，均無須再為文法規則傷透腦筋。學英文時首重建立 chunks、set-phrases 和 word partnerships 語庫，多學多益，你再也不必勞心費神學文法，或者苦苦思索如何在文法中套入單字。這三類 MWI 用來輕而易舉，而且更符合人腦記憶和使用語言的習慣。本節結束前，請做最後一個 Task，確定你對 Leximodel 已經完全了解，同時驗證這方法確實簡單好用。

Task 6

請看以下電子郵件和摘要，找出 chunks、set-phrases 和 word partnerships，並分別用三種不同的顏色畫底線，最後完成下表。見範例。

Dear Marge,

Thanks for your email, which I received yesterday. I have been very busy so I have not had time to look for the media review data you asked for. I hope I'll have time to get it to you tomorrow. Regarding the customer satisfaction survey, this will be ready next week. In your email you mentioned that you are worried about the focus groups. Don't worry! We are currently organizing some people for the focus group to test the TV commercial and have already booked the research company for Saturday 28th at 9:00. I saw the draft version at the production company last night and the result looks good.

If you have any more questions, please don't hesitate to call me. I'm looking forward to seeing you at the focus group on Saturday.

Regards,
Oliver

♀ Email 摘要

奧立弗和瑪姬為廣告界人士。奧立弗寫信向瑪姬致歉，表示他一直很忙，因此未將媒體評估報告寄去。他向瑪姬報告案子進度：客戶滿意度調查即將完成，他已向調查公司預約，並找到焦點團體測試公司製作的廣告。看過廣告初版，他覺得做得還不錯。

Set-phrases	Chunks	Word Partnerships
Thanks for your email.	... be ready ...	research company

請以下列語庫核對你的答案。

Email 必備語庫　前言 3

Set-phrases

- Thanks for your email.
- Regarding n.p. ...
- In your email you mentioned that + n. clause ...
- If you have any more questions, ...
- ..., please don't hesitate to V ...
- I'm looking forward to Ving

Chunks

• ... have been very busy ...	• ... be worried about ...
• ... have time to V ...	• ... are currently organizing ...
• ... look for ...	• ... have already booked ...
• ... get sth. to sb. ...	• ... book sth. for + date ...
• ... asked for sth. ...	• ... see sth. ...
• ... be ready ...	

Word Partnerships

• customer satisfaction survey	• research company
• media review data	• draft version
• focus group	• production company
• TV commercial	• last night
	• looks good

💡 語庫小叮嚀

- 注意 set-phrases 通常以大寫開頭，或者以句號結尾。三個點（...）表示句子的流動部分。
- 注意 chunks 的前後都有三個點，表示 chunks 大多為句子的中間部分。
- 注意所有的 word partnerships 都至少包含兩個意義字。

如果你的答案不若參考答案完整，別擔心。要能找出一篇文章中所有的固定元素，需要多加練習才行。但是我保證一旦你能找出像參考答案一樣多的 MWI，那就表示你的英文已經達到登峰造極的境界了！很快你便能擁有這樣的能力。於本書末尾，我會請你再做一次這個 Task，讓你自行判斷學習成果。現在有時間的話，建議你找一篇英文文章，英文母語人士所寫的電子郵件、雜誌或網路上的文章皆可，請以該篇文章做相同的練習。

本書架構與使用說明

◐ 本書的架構為何？

　　本書包含兩個 Section，區分為「學習篇」與「工具篇」。Section 1「學習篇」共分七個 Unit（為中文解說方便，以下以「章」稱之。），每章鉅細靡遺地教你電子郵件的一種寫法或類型。例如第二和第七章教的是電子郵件的開頭和結語。第三至第五章教的是三種最常見的電子郵件類型——請求信、安排會面信和解決問題信。第六章教的是電子郵件的回信程序和寫法，並以資訊交流和有效率的商業溝通為重點。現在請花一點時間看看本書目錄，熟悉接下來要學習的重點。

　　每一章都有三節。第一和第二節會講解 set-phrases；第三節則是「語感甦活區」，教學重點為 chunks 或 word partnerships，這兩者是以中文為母語的專業人士在使用英文時常遇到的問題。你會發現「語感甦活區」對於電子郵件寫作或加強英文有幫助。但如果你的時間不多，可以先跳過第三節不看，但是以後有時間時還是建議你回頭看完。此外，每一章均有多個電子郵件範例，為你增加接觸好範文的機會。本書的電子郵件範文沒有完整中譯，因為中文電子郵件的格式和英文郵件的格式可謂南轅北轍，我不希望你耗時看翻譯。但是為了幫助理解，本書會提供中文摘要。請先盡量看懂英文電子郵件，再看中文摘要。

　　Section 2「工具篇」則分為九章，依職場工作者實務上最常遇到的情況，如『回信的開場白』、『安排會面』、『急件』、『結語』等，分門別類歸納出好用句型及例句，依主題查找到相關句子之後照抄即可。

　　除此之外，書末附錄嚴選了各章節共計約 40 類必備語庫，並列出高頻字串及便利句，方便急需時迅速助你一臂之力。

◎ 我該如何實際運用 Leximodel 學英文？為什麼 Leximodel 和我以前碰到的英文教學法截然不同？

簡而言之，我的答案是：只要知道字詞的組合和這些組合的固定程度，就能簡化英語學習的過程，同時大幅減少犯錯的機率。

以前的教學法教你學好文法，然後套用句子，邊寫邊造句。用這方法寫作不僅有如牛步，而且稍不小心便錯誤百出，想必你早就有切身的體驗。現在只要用 Leximodel 建立 chunks、set-phrases 和 word partnerships 語庫，接著只須背起來就能學會英文寫作了。

◎ 這本書如何以 Leximodel 教學？

本書提供電子郵件中所有常見的固定字串（chunks、set-phrases 和 word partnerships，但以 set-phrases 最多），告訴你如何記憶和運用這些字串，並教你如何每天留意並記下所看到的英文，幫助增強英文基礎。

◎ 為什麼要留意字串中所有的字，很重要嗎？

不知何故，大多數人對眼前的英文視而不見，分明擺在面前仍然視若無睹。他們緊盯著字詞的意思，卻忽略了傳達字詞意思的方法。每天瀏覽的固定 MWI 多不勝數，卻未發覺這些 MWI 是固定、反覆出現的字串罷了。任何語言都有這種現象。這樣吧，我們來做個實驗，你就知道我說的是真是假。請做下面的 Task。

🔍 Task 7

請由下列選出正確的 set-phrases。

- Regarding the report you sent me ...
- Regarding to the report you sent me ...
- Regards to the report you sent me ...
- With regards the report you sent ...
- To regard the report you sent me ...
- Regard to the report you sent me ...

　　姑且不論答案為何，我敢說這題很難，是吧？每天你大概都會看到這個 set-phrase，卻從未刻意地仔細看它當中的每一個字。（**其實第一個 set-phrase 是正確答案，其他一概錯誤！**）這就證明了我在教 set-phrase 時提到的第一個小叮嚀沒錯。無論如何，一定要加強注意所接觸到的文字，但**更重要的是確定當作範本的電子郵件百分之百是出自英文母語人士之手。**所謂的「英文母語人士」，我指的是美國人、英國人、澳洲人、紐西蘭人、加拿大人或南非人，其他人的電子郵件都不夠可靠。如果英文非母語，就算是老闆寫的電子郵件也不可完全信任。公司中若有人在十年前到美國唸過博士，英文能力公認好得沒話說，他們的電子郵件也信不過。

　　如果多留意每天接觸到的固定字串，久而久之一定會背起來，轉化成自己英文基礎的一部分，這可是諸多文獻可考的事實。刻意注意閱讀時遇到的 MWI，亦可增加學習效率。Leximodel 正能幫你達到這一點。

◉ 須小心哪些問題？

　　本書中許多 Task 的目的，即在於幫你克服學 set-phrases 時遇到的問題。學 set-phrases 的要領在於：**務必留意 set-phrases 中所有的字。**

　　從 Task 7，你已發現其實自己不如想像中那麼留意 set-phrases 中所有的字。接下來我要更確切地告訴你學 set-phrases 時的注意事項，這對學習非常重要，請勿草率閱讀。一般而言，學習 set-phrases 時容易出錯的問題，主要可歸納成以下三類：

1. 短字

（如 a、the、to、in、at、on 和 but）這些字很難記，但是瞭解了這點，即可說是跨出一大步了。Set-phrases 極爲固定，用錯一個短字，整個 set-phrase 都會改變，等於是寫錯了。

2. 字尾

（有些字的字尾是 -ed，有些是 -ing，有些是 -ment，有些是 -s，或者沒有 -s。）字尾改變了，字的意思也會隨之改變。Set-phrases 極爲固定，寫錯其中一字的字尾，整個 set-phrase 都會改變，等於是寫錯了。

3. Set-phrases 的結尾

（有的 set-phrase 以 n. clause 結尾，有的以 n.p. 結尾，有的以 V 結尾，有的以 Ving 結尾，我們稱之爲 codes〔分類碼〕。）許多人犯錯，問題即出在句子中 set-phrase 與其他部分的銜接之處。學習 set-phrases 時，必須將 code 當作 set-phrases 的一部分一併背起來。Set-phrases 極爲固定，code 認錯了，整個 set-phrase 都會改變，等於是寫錯了。

　　教學到此，請再做一個 Task，確定你能夠掌握 codes 的用法。

Task 8

請看以下 codes 的定義，然後按表格將字串分門別類。第一個字串已先替你找到它的位置了。

- **n. clause** = noun clause（名詞子句）
  n. clause 一定包含主詞和動詞。例如：I <u>need</u> your help.、She <u>is</u> on leave.、We <u>are</u> closing the department.、What <u>is your</u> estimate? 等。
- **n.p.** = noun phrase（名詞片語）
  這其實就是 word partnership，只是不含動詞或主詞。例如：financial news、cost reduction、media review data、joint stock company 等。

• **V** = verb（動詞）

V 和 n. clause 不同，因為 V 不連接主詞。

• **Ving** = verb ending in -ing（動詞以 -ing 結尾）

以前你的老師可能稱之為名詞，但其實這是不對的，Ving 僅是看似名詞而已。

bill of lading	having	our market share
customer complaint	he is not	your new client
decide	help	see
did you remember	helping	sending
do	I'm having a meeting	talking
doing	John wants to see you	we need some more data
go	knowing	wrong figures
great presentation	look after	you may remember

n. clause	n.p.	V	Ving
	bill of lading		

Task 8 參考答案

請以下列語庫核對你的答案。

n. clause	n.p.	V	Ving
you may remember	wrong figures	help	helping
we need some more data	customer complaint	do	knowing
did you remember	bill of lading	see	doing
John wants to see you	our market share	look after	having
I'm having a meeting	your new client	decide	sending
he is not	great presentation	go	talking

💡 **語庫小叮嚀**

• 注意 n. clause 的 verb 前面一定加主詞。

• 基本上 n.p. 即為 word partnerships。

◎ 如果沒有文法規則可循，我該如何知道自己的 set-phrases 用法正確無誤？

為了助你一臂之力學好英文，本書提供了 language banks（語庫）。你只須檢查自己在電子郵件中用到的 set-phrases，和本書語庫的一模一樣即可。張大雙眼，用心看我上面講述的所有細節準沒錯。別擔心為何你用的 set-phrases 不對，或者運用或違反了哪些文法規則，參閱本書的語庫就沒問題。

🔍 **Task 9**

請看下列句子中的錯誤，並與前言語庫 3 的 set-phrases 作比較。在題號前寫下錯誤原因的編號（1. 短字；2. 字尾；3. set-phrases 的結尾）。見範例①。

____2____ ① Thank for your email.

_____ ② Regarding you want to see me, I can come to your office on Monday.

_____ ③ On your email you mentioned that you are worried.

_____ ④ If you have any more question, please call me.

_____ ⑤ If you need help, please don't hesitate to contacting me.

_____ ⑥ I'm looking forward seeing you on Monday.

_____ ⑦ I'm looking forward to see you again.

Task 9 參考答案

② 3   ③ 1   ④ 2   ⑤ 3   ⑥ 1   ⑦ 3

Task 10

現在請改正並寫出正確的句子，然後再檢查答案。

① Thank for your email.

→ *Thanks for your email.* _____

② Regarding you want to see me, I can come to your office on Monday.

→ _____

③ On your email you mentioned that you are worried.

→ _____

④ If you have any more question, please call me.

→ _____

⑤ If you need help, please don't hesitate to contacting me.

→ _____

⑥ I'm looking forward seeing you on Monday.

→ _____

⑦ I'm looking forward to see you again.

→ _____

② Regarding <u>our appointment</u>, I can come to your office on Monday.

③ <u>In</u> your email you mentioned that you are worried.

④ If you have any more <u>questions</u>, please call me.

⑤ If you need help, please don't hesitate to <u>contact</u> me.

⑥ I'm looking forward <u>to</u> seeing you on Monday.

⑦ I'm looking forward to <u>seeing</u> you again.

　　如果你的答案和參考答案南轅北轍，請再重新看看本節，並特別留意 Tasks 以及對於 set-phrases 三個細節的解說。另外也可再看一次 Task 6 的電子郵件，參考其中 set-phrases 的用法。如果有必要，請現在就回頭複習。

　　本書中許多 Task 會幫你把注意力集中在 set-phrases 的類似細節上，你只須作答和核對答案，無須擔心背後原因。

◉ 我該如何利用本書達到最佳學習效果？

　　為了達到最佳學習效果，謹提供以下幾個學習訣竅：

一、請逐章逐節看完本書。為了提供更多記憶 set-phrases 的機會，本書會反覆提到一些語言和概念，因此倘若一開始有不解之處，請耐心看下去，多半讀到本書後面的章節時自然就會恍然大悟。

二、請將每天所寫的電子郵件列印出來，每看完本書的一章就取出閱讀，看看哪裡有錯，該如何改正。

三、每個 Task 都要做。這些 Task 有助於記憶本書中的字串，亦可加強你對這些字串的理解，不可忽視。建議使用鉛筆做 Task。如此寫錯了還可以擦掉再試一次。

四、每隔幾星期溫習相關章節中所有的語庫，以免忘記字串的用法。

五、做分類 Task 時（見第二章 Task 2.2），在每個 set-phrase 旁做記號或寫下英文字母即可。但是建議有空時，還是將 set-phrases 抄在正確的一欄中。還記得當初是怎麼學中文的嗎？抄寫能夠加深印象！

六、每星期花二十分鐘閱讀一個電子郵件範例，找出所有 set-phrases 和 chunks、畫底線，並將其中一些陌生的字串抄寫幾遍，寫信時嘗試使用這些字串。花在抄錄字串的時間絕對不會白費。

七、請記錄字串的使用狀況，寫下平日電子郵件中會使用的字串。

八、如果提升英文能力的決心堅定，建議和同學、同事、親朋好友組成讀書會，一同閱讀本書和做 Task。

　　當你讀完本書，做完所有習題、運用本書的字串練習寫電子郵件，也時時比較現在和以前所寫的電子郵件之後，你的英文電子郵件應該已經大有進步。但是接下來要如何確保不忘所學，鞏固已有的英文水平？許多客戶告訴我，一旦離開英語的生活或工作環境，便很難維持或提升英文程度。在此我提出一些建議，告訴你持續加強英文的好方法。

◔ 如何維持英文水平

　　保持英文能力的方法不計其數，我把建議的方法劃分為六大類如下：

一、每天背幾個新字串

你可以從報紙、網上或收到的電子郵件（別忘了挑選英文母語人士所寫的電子郵件當範本）中挑出字串。切記在字串上下功夫，也就是 chunks、set-phrases 或 word partnerships，若只是一昧地背單字，對學英文毫無意義。

二、選擇難、陌生或新的字串

記住，天底下沒有難的字串，只有不熟悉的字串。有些字串乍看之下覺得陌生，但是只要多加練習，便會變得熟悉。新字串一旦開始運用自如，就表示你已經學會，可以繼續學習下一個新字串了。

三、刻意運用新字串

利用本書的附錄，嘗試運用其中的字串，有益於記憶和吸收，也可將其轉化成自己詞彙的一部分。

四、如果可以，盡量避免使用已知和運用自如的字串。

許多人總是反覆使用同樣、已知、運用自如的字串，也因為如此，他們的英文永遠停留在原地，不會進步！

五、試用不同用法，勇於嘗試，從錯誤中學習。

用字遣詞會出錯，是因為試圖發揮創意，嘗試不同的用法，這是學習語言過程中很重要的一環，不容小覷。從不犯錯，並不代表你很優秀（除了極少數的例外），實際上你只是裹足不前，不敢嘗試新語文！

六、留意生活週遭遇到的語文

嘗試創造一個屬於自己的英文環境。從把喜歡的網站加到「我的最愛」名單中開始，閱讀感興趣的文章，花些時間將其中的 chunks 和 word partnerships 抄下來。研究一再證明，加強英文最重要的方式毫無疑問地就是閱讀、閱讀、再閱讀！

◎ 在開始研讀本書之前，還有哪些須知？

Yes, you can do it! 不要擔心！只要依循本書架構，循序漸進地學習，一定能寫好英文電子郵件。

翻開第一章前，請回到前言的「學習目標」，回顧一下並勾出自認為達成的項目。希望全部都能夠打勾，如果沒有，請重新閱讀相關段落。

Section 1
學習篇：基礎字串概念

Unit 1
商務 Email 寫作的基本要領

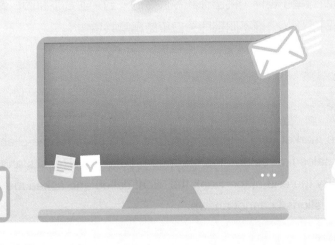

　　本章將解說＜前言＞中的第四個問題。還記得嗎？那個問題是：電子郵件的風格為何？

　　許多人對這個問題感到疑惑，但是我認為是因為很多人對電子郵件寫法的認知有誤，總是深怕遺漏了某些重要資訊。或許你認為電子郵件和商業書信一樣，後者有特定的風格和正式規格，前者也就必須遵循特定規則和標準，但實則不然。也或許你會擔心語氣有欠禮貌或過度直接，甚至長度不恰當，這其實是多慮了。本章中，我將會告訴你寫商務電子郵件的原則，幫你一掃疑慮，並有本事判斷電子郵件寫得是優是劣。為了讓你了解這些問題，我們先做個 Task 吧。

Task 1.1

請閱讀下列四封電子郵件，挑出你認為最好的一封，並寫下原因。喬治和瑪莉在同一家公司工作，分屬不同部門，但平時經常共事。

①

> Hi George,
>
> I need the market data. Please send me the figures for the client. I also need the sales figures and the revenue breakdown. Results and targets together. By the end of the week the client called and asked me a whole load of questions I cannot answer.
>
> Please help.
>
> Thanks,

Vocabulary

figure [ˈfɪgjə] *n.* 數值；數據　　**breakdown** [ˈbrekˌdaʊn] *n.* 明細

②

Dear Mr. George,

Following my aforementioned telephonic communication with our esteemed and estimable client in the Middle Kingdom, who was concerned about the status of her commercial concerns in that aforementioned place, I would be extremely grateful and eternally obliged to you for your great kindness and the honor you do me if you could allow me to receive before the passage of much time has passed the figures I need to fully answer the clients concerns and put her mind at rest so that we can facilitate a friendly and smooth flow of business between my office and hers, thus allowing all of us to benefit in a fiscal — and indeed a humanitarian sense ...

③

Dear George,

Client called you to me ask for that I need a figure to tell client: Market share data for sales and budget results could you?

Thanks,

④

Dear George,

Client XYZ has just called asking for an update on the situation in the China market. So that I can answer her questions more fully, could you send me the following figures which we discussed at our last meeting?

The market share data

Vocabulary

client [`klaɪənt] *n.* 客戶；委託人

fiscal [`fɪskl] *adj.* 財政的；會計的

The sales data, both targets and results
The revenue and cost breakdown

I'll need these by the end of the week.

Thanks,
Mary

⚡ Email 摘要

瑪莉寫信向喬治索取資料，以便轉交給一名客戶。她向喬治索取中國市場佔有率的數據、業績目標和成果，以及最新收益和成本明細表。這禮拜必須寄給她。

📖 Task 1.1 答案

第四封電子郵件寫得最好。

能夠選出最好的一封電子郵件固然重要，但知道該封信件好在哪裡卻更重要。你要根據哪些準則做判斷呢？

📖 Task 1.2

以下為判斷電子郵件好壞的準則，哪一些最接近你的標準，請打 ✓。

(　) 表達清楚 　　(　) 語氣正式 　　(　) 詞句準確
(　) 長度恰當 　　(　) 文筆優美 　　(　) 資訊完整

📖 Task 1.2 答案

正確答案是**表達清楚**和**資訊完整**。

在 Task 1.1 中，第四封電子郵件寫得最好，因為要求的事項最清楚，簡明扼要且有條不紊，詞句也正確無誤，讀起來一目了然。希望你答對了！

一封電子郵件寫得是優是劣，「清楚明白」應該是最重要的準則。用字遣詞和內容長度則視需要而定，例如：是否有為對方解釋的必要，讓對方看了郵件後就知道該怎麼做。換句話說，傳達訊息永遠是寫電子郵件時的重點。本章將更加詳細地說明「清楚明白」這個原則。請先看一下本章學習目標的概要，然後繼續往下閱讀第一節。研讀完本章，你應該達到的學習目標如下：

☐ 清楚知道寫電子郵件時，最重要的是抓住訊息重點。
☐ 收集電子郵件最常見的訊息類型。
☐ 了解寫電子郵件時，文字必須一目了然。
☐ 了解寫電子郵件時，詞句必須正確，以及如何達到此原則。
☐ 了解何謂簡明扼要的電子郵件，以及此原則的重要性。
☐ 了解如何應用上述原則寫電子郵件。
☐ 練習應用這些原則寫電子郵件。

◑ 讓收件者優先處理的電子郵件，關鍵就在於主旨！

「清楚明白」這個準則指的不只是訊息內容而已，語意清晰的主旨也扮演了重要角色，因為它對收件者有以下兩大功能：

一、準確掌握來信的內容，以及此封郵件是否必須回覆。

收件者一天會收到許多垃圾郵件，而他們經常連看都不看就把那些信刪除掉。一個清楚的主旨將有助於收件者知道你的來信相當重要。

二、整理收件匣，使其有條不紊。

收件者能夠透過清楚的主旨，迅速找到郵件，節省時間以提高辦公效率。

撰寫主旨時，你應該使用聚焦於待辦業務的 word partnership；切勿寫出一長串完整的句子，**用一個動詞、名詞或形容詞的詞組即可**。以下我們來看一些例子。請看第 45-47 頁的十封電子郵件訊息，試著將下列主旨配對到正確的電子郵件，見範例。

	Ⓐ Account number XYZ 123		Ⓕ Package confirmation
1	Ⓑ Appointment cancellation		Ⓖ Project postponement
	Ⓒ File enquiry		Ⓗ XYZ project meeting
	Ⓓ IT reminder		Ⓘ Your file
	Ⓔ New product line enquiry		Ⓙ Your report

答案

Ⓐ10　　Ⓒ8　　Ⓓ6　　Ⓔ5　　Ⓕ7　　Ⓖ4　　Ⓗ9　　Ⓘ3　　Ⓙ2

訊息清楚明白

剛剛已經說過，寫商務電子郵件時，最重要的考量就是將業務訊息寫得愈清楚愈好。接下來看一看幾種最常見的電子郵件。首先請做以下的 Task。

Task 1.3

請閱讀以下喬依絲所寫的電子郵件，然後在第 48 頁的表中填入適當的郵件編號。見範例。

①

Dear William,

I'm afraid I have to cancel our appointment on Friday as I have to go to HK on urgent business. I apologize for the short notice. I'll be in touch when I get back to fix a new meeting time. Sorry once again, and thanks for your understanding.

②

Dear Judy,

Thank you for sending the report so quickly. I have had a brief look through it and it looks fine. I hope to have a chance to look through it in more detail over the next week.

Vocabulary

look through 仔細閱讀

③

Dear Max,

I have attached the file you requested for your reference. I apologize for the delay in getting back to you, but I have been on leave. Please contact me if you have any further questions about this.

④

Dear Oliver,

Due to the merger, we are having to postpone all projects until the middle of next year. We apologize for this, but hope you will understand.

⑤

Dear Joseph,

The client has written to me asking for details on the new product line. As I do not have the specs on hand, do you think you could deal with her query? I've included her emails and contact details below. If you have any questions about it, please contact me.

⑥

Dear All,

Please remember to exit all programs before shutting down your workstations at the end of the day. Turning off your PC without exiting all programs causes big problems for the IT staff when starting up the systems the next day. Your help is appreciated.

Vocabulary

spec [spɛk] *n.* 明細；規格（specification [ˌspɛsəfəˈkeʃən] 之縮寫）

⑦

Dear John,

Could you please confirm whether the package I couriered over to you on Wednesday has arrived or not? I haven't heard from you about it and I am worried. Thanks.

⑧

Dear Mark,

Thanks for sending the file, which I have looked at carefully. However, I have a few questions. Could you explain why the chart on page 15 only shows the figures for the North region and not all the regions? Also, do you think you could let me know whether you have included our loss from last year in your projections? You do not really make this clear.

⑨

Dear Lulu,

I was wondering if we could meet up to discuss the XYZ project that we will be working on together. Would next Wednesday at 10 am in my office be OK for you?

⑩

Dear Jojo,

Thanks for your email. Regarding your query about account number XYZ 123, the file is stored in clients. I hope this is clear.

Vocabulary

courier [ˈkʊrɪə] *v.* 快遞   **query** [ˈkwɪrɪ] *n.* 詢問

請作答：

arranging a meeting time	9	requesting action	
cancelling a project		requesting confirmation	
changing an arrangement		requesting information	
confirming receipt		responding to a request	
reminding		sending information	

Task 1.3 參考答案

請閱讀下列電子郵件摘要，即可看出答案為何。

♥ Email 摘要

1. 喬依絲要更改和威廉原訂的會議時間。她得到香港出差，所以會議必須取消，並向威廉表示歉意。
2. 喬依絲向茱蒂確認收到報告，她會在下星期仔細閱讀。
3. 喬依絲寄出麥克斯索取的資訊。她因為請假拖延了此事，在信中向麥克斯解釋原因並表示抱歉。
4. 喬依絲的公司和其他公司合併，因此案子延期了。
5. 喬依絲請約瑟夫幫忙處理事情。她請約瑟夫回覆一名客戶的問題，因為她手邊沒有相關資料，但是他有。
6. 喬依絲提醒員工，下班關電腦前別忘了先結束使用中的程式，避免製造資訊部人員的麻煩。
7. 喬依絲向約翰確認是否收到快遞送去的包裹。
8. 喬依絲向馬克詢問報告的細部問題。她要知道為何第 15 頁的圖表不完整，還有馬克的預測是否包含去年的虧損。
9. 喬依絲和盧露敲時間討論 XYZ 專案，提議下星期三 10 點在她辦公室討論。
10. 喬依絲回信給邱裘。對方向她索取帳戶資料。

從本章開始，本書將針對以上常見的電子郵件類型，教你寫出好郵件。在此我首先教的是 transparency（清楚明白）。基本上，這個原則的概念就是**電子郵件必須清楚明白，有效地傳達訊息**。也就是說，**語氣不要太過正式，內容也不要有錯誤，以免帶給對方突兀感**。請做下面幾個 Task，你就會了解我的意思了。

Task 1.4

請看以下這封電子郵件。這是哪一種訊息類型、所要傳達的訊息為何？請寫下這封郵件訊息不清楚的原因。

Dear Joseph,

Following our extremely useful and entertaining telephonic communication on the 3rd day of this month in the year of our Lord Anno Domini 2014, I would respectfully wish to draw your attention to the digital file which I have previously attached to this message.

During our discourse, we conversed about the need for me to provide you with a more profound and deeper understanding of the many issues involved with bringing to fruition the project in which we are all involved to the mutual financial benefit of all parties concerned in the enterprise. You mentioned specifically the need for fuller statistical data. I sincerely hope that I will not be contradicted if I maintain that the attached file should go some way towards meeting this requirement. If, however, you find that there are areas of incompleteness or – Heaven forefend and forgive me for saying so! – unclear, I hope you will not think uncharitably of me but will contact me and request clarification on those issues which you currently find obfuscating.

I remain your obedient servant,

Vocabulary

discourse [ˈdɪskors] *n.* 談話

下列何者為上封郵件所要傳達的訊息？

① 寄件者想當約瑟夫的傭人。

② 寄件者向上帝祈禱。

③ 寄件者在報告他和約瑟夫的談話。

④ 寄件者寄出約瑟夫先前索取的文件。

⑤ 寄件者和約瑟夫吵架。

Task 1.4　參考答案

其實④才是正確答案。答錯了別沮喪，這個 Task 的目的是為了說明，這封電子郵件的用字遣詞艱澀，閱讀時，你的注意力放在看懂字詞的意義，而非訊息本身，因此訊息不夠明確。

上封郵件語意不明的三大原因：

一、句子冗長，結構複雜。

二、無關的資訊太多。

三、很多字串意義模糊、過於老式，一般很少用於商務電子郵件中。

由此得知，寫商務電子郵件時應謹記以下三點，方可清楚傳達訊息：

一、句子短而簡單。

二、不寫無關的資訊。

三、使用自己領域的商務電子郵件中常見的字串。

下面的 Task 可加深你對這些原則的印象。這個 Task 有點難，但是千萬別偷懶不做。做了 Task 後，你便會知道如何寫出言詞達意的電子郵件，也有助於理解本書接下來的教學重點。

Task 1.5

請改寫 Task 1.4 的電子郵件，使訊息清楚明白。做完 Task 前，不要先看參考答案。

Dear Joseph,

Further to our call on the 3rd, I have attached a file containing the data for the project as you requested.

If you have any questions about it, please contact me.

希望你寫的電子郵件如參考答案般簡明扼要。如果沒有，別擔心，本書在接下來的章節中會教參考答案中的字串，等你看完全書再回頭做這個 Task 時，應該就會覺得易如反掌了。

✔ 訊息正確

電子郵件的訊息不清楚，另一個原因是詞句有錯，簡單地說就是文法和單字出錯，導致語意不詳。請看下面 Task 中的範例。

Task 1.6

請閱讀以下電子郵件。這些電子郵件要傳達的訊息為何？

①

Dear Marge,

Please notice the new client when you are moving into your new office. Please send it your new contact details.

②

Dear Roli,

due to the timing concerns so that Tracey follow PSQ report already booked the care label accordingly! as long as follow PSQ report which we feel comfortable. Thanks for your understanding!

Task 1.6　參考答案

老實說，很難猜出這兩封電子郵件的訊息為何。電子郵件①中，我猜寄件者是請瑪姬 notify（告知）而非 notice（注意）客戶，然後要瑪姬將聯絡方式寄給客戶。但是因為錯將 notify 寫成 notice，them 也寫成 it，讓人一頭霧水。郵件②更是不知所云，如果收到這封信，八成得回信請對方重新解釋了。

電子郵件中用詞錯誤，十之八九是因為郵件中的 chunks 和 set-phrases 不完整或出錯。還記得＜前言＞中關於 set-phrases 細節的 Task 嗎？使用 chunks 時同樣得注意細節。請做下面的 Task，你便會了解使用 set-phrases 和 chunks 時必須一字不差的重要性了。

 這個 Task 請重複做兩次，第一次先不要看下方的語庫，第二次再借助語庫做 Task。

Task 1.7

請更正下列電子郵件中的錯誤。若有必要，可參考下方的語庫。

Dear Marge,

Thank for you email, which I receive yesterday. I have very busy so I have not time look the media review data you asked. I hope I'll time to get to tomorrow.

Regarding which the customer satisfaction survey, this will be next week. In email you mentioned worry the focus groups. Don't worry!

Organizing some people for the focus group to test the TV commercial and did already book the Saturday 28th at 9 research house. I saw at the production house last night and the result good.

If you have any more questions, please hesitate call.

I'm look forward to see at the focus group on Saturday.

Regards,
Oliver

Set-phrases	Chunks
● Thanks for your email ...	● ... have been very busy ...
● Regarding n.p. ...	● ... have time to V ...
● In your email you mentioned that + n. clause ...	● ... look for ...
	● ... get sth. to sb. ...
● If you have any more questions, ...	● ... asked for sth. ...
● ..., please don't hesitate to V ...	● ... be ready ...
● I'm looking forward to Ving	● ... are currently organizing ...
	● ... have already booked ...
	● ... book sth. for + date ...
	● ... see sth. ...

Task 1.7 參考答案

請將答案與＜前言＞中 Task 6 的電子郵件作比較。答案正確固然重要，但是我更希望你能看出這些錯誤都是 chunks 和 set-phrases 不完整或用錯所致。

◎ **訊息簡明扼要**

現在我要教你寫電子郵件時的最後一個原則，也就是 concision（簡明扼要）。這個原則較難解釋，也較難理解，但與中文翻譯成英文的習慣有關。我來舉一個例子，請比較下面意義相同的中英文句子。

I want to get off the bus.
我想下公車。

以中文來說，這個句子需要五個字來表達，而英文需要七個字。但是，有時中文需要用的字反而比較多。請看以下例子：

My brother is reading.
我弟弟（或哥哥）在看書。

有時英文電子郵件的詞意不清，問題只是在於使用的字不夠，無法完全表達意思。以下提供幾個簡易的原則，幫你避免這種問題：

一、每個動詞都要有一個主詞。
二、每個句子都有完整的句意。
三、以 chunks 和 set-phrases 為單位來翻譯，而非以單字。

Unit
1

要避免寫出不夠扼要的電子郵件，最重要的就是從 Leximodel 的角度看語言：句子是由字串構成，而非一個個的單字。現在來做最後一個 Task，看看簡明扼要的郵件怎麼寫。

Task 1.8

請閱讀下列電子郵件，並更正其中的錯誤，然後判斷各錯誤是違反了上述哪一個原則。

> Dear Jordan,
>
> My boss go last weekend Hualien. Very happy and have a lot of interest towards the project please could send more information thanks about the market share.

Task 1.8　參考答案

> Dear Jordan,
>
> My boss went to Hualien last weekend. She is very happy and is very interested in the project. Could you please send us more information about the market share?

親愛的喬登：

我老闆上星期去花蓮，很愉快，她對案子有高度興趣，煩請寄更多市場佔有率的相關資料給我。

希望你在比較上列電子郵件的三個版本（兩個英文版本、一個中文版本）之後，能看出錯誤大多是因為逐字翻譯、用中文書寫習慣來寫英文所致。照這種方法永遠寫不出正確的英文，唯一的解決之道就是按照本書的指導，學會正確使用 chunks 和 set-phrases。

在本節中，我要介紹些電子郵件中最實用的字串。以下請做 Task。

Task 1.9

請從表格下方找出適當動詞，填入下表與各名詞配對成正確詞組。各名詞可搭配多個動詞，動詞也可重複使用。見範例。

send / clarify	information		an arrangement
	a suggestion		a procedure
	an invitation		a plan
	a query		a request
	a process		an instruction
	a proposal		a question
	an offer		background
	a complaint		a reason
	action		a due date
	a reference to		an explanation
	an order		

- accept
- announce
- cancel
- change
- clarify
- confirm
- confirm receipt of
- convey
- deal with

- demand
- describe
- follow up
- give
- implement
- initiate
- make
- offer
- outline

- place
- put forward
- put into action
- reject
- request
- respond to
- send
- set
- settle

請研讀下列語庫，並檢查答案。

Email 必備語庫 1.2

send / request / convey / clarify	information
confirm / describe / make	an arrangement
make / offer / reject / respond to / accept	a suggestion
describe / outline / initiate	a procedure
make / offer / reject / respond to / accept	an invitation
outline / put forward / put into action / implement / announce	a plan
respond to	a query
make / respond to / deal with	a request
describe	a process
give / cancel / change	an instruction
make / offer / reject / respond to / accept	a proposal
raise / ask / answer / respond to	a question
make / reject / respond to / accept	an offer
describe / outline	background
make / deal with / settle	a complaint
give	a reason
request / confirm / follow up	action
give / set	a due date
make	a reference to
give / request / demand	an explanation
change / cancel / place	an order

※ 延伸閱讀看例句：〈附錄二〉（P. 354）

Task 1.10

請研讀語庫 1.2，然後練習以語庫中的字串造句。

Task 1.10　參考答案

- I need to **place an** urgent **order**.
  我要下訂單，急件。

- Could you please **deal with this complaint** from a customer?
  能否請你處理這位客戶的投訴？

- I think we should **put the plan into action** immediately.
  我認為我們應該立即著手進行計畫。

- We need to **initiate** firmer safety **procedures**.
  我們必須啓用更穩當的安全程序。

- Please can you **give a reason** for the price increase?
  可不可以麻煩你提出漲價的理由？

- My customers are **demanding an explanation** for the faulty product.
  對於產品出錯，我的顧客要一個解釋。

- I will **make arrangements** to have someone meet you at the airport.
  我會安排找人去機場接你。

- So far you have still not **responded to** my **query**.
  到目前為止你還沒有回答我的疑問。

　　最後，本章結束前，建議你看一看最近寫的電子郵件。仔細閱讀全文，同時思考本書提到的「清楚明白」、「正確」和「簡明扼要」等原則。有哪些地方需要改進？該如何改進？現在請花點時間照做，並作筆記，你會發現這對學習相當有助益。

　　好啦，總算進入本章尾聲，結束前，請回顧本章前言中所設定的學習目標，勾出自認為已達成的項目。如果還有不懂的項目，請在幾天後重新閱讀相關章節，相信第二次閱讀時你便會了解。

Unit 2
開場白
Set-phrases

　　電子郵件大致上可分為兩大類：發信和回信。通常從信件開端可看出電子郵件屬於哪一類信件，同時得知所交代的事情為何，也就是說，信件的開端有說明信件主旨的作用。一般說來，電子郵件在開端的用語也較制式，不妨將這些用語當作 set-phrases 來學習和使用。我們先來看看，下面幾封電子郵件屬於哪一類信件？現在請做 Task 2.1，然後檢查答案。

Task 2.1

以下電子郵件是發信（Initiating）還是回信（Responding）？請填寫於空格中。你從哪些字串看出端倪？請在該字串下畫線。見範例 ①。

① Dear George,

Thanks for your email about the meeting on Friday. I'm sorry I was not able to attend, but my daughter had to go to the dentist for emergency surgery and my wife is away on business.

　　　Responding

② Dear Kevin,

Thanks for your message, and apologies for the delay in getting back to you. Things have been quite busy here with the merger. Regarding your inquiry, I am meeting the CEO tomorrow afternoon and will put your question to him. I'll let you know what he says tomorrow afternoon, OK?

Vocabulary

merger [ˈmɝdʒɚ] *n.* 合併

③ Dear Louis,

Just to remind you that all monthly accounts must be completed by the 28th, and not by the last Friday of the month as before. Thanks.

---

④ Dear Macey,

Just to confirm our appointment on Friday. Should I bring the dossier to the meeting or would you like me to send it over beforehand by courier?

---

⑤ Dear Shirley,

Thank you for sending your resume to me. I'm afraid we do not currently have a position for someone of your experience. However, I will keep your details on hand and get back to you if such a position does become available.

💡 **Email 摘要**
1. 寄件者謝謝喬治告知他會議的事，並向他道歉。他女兒必須去看牙科，因此不克參加會議。
2. 寄件者向凱文道歉。並向凱文解釋她的公司正在進行合併計畫，因此未能儘速回信答覆凱文的問題。她告訴凱文，她將在會議上幫他詢問執行長，下午再給他答覆。
3. 寄件者提醒路易士，每月結單的繳交期限有變動。
4. 寄件者跟梅西確認見面時間，並問她是否希望事先拿到資料文件。
5. 寄件者告知雪莉公司目前沒有職缺，但將保留她的履歷表和資料，以便日後有職缺時通知。

Vocabulary

dossier [ˈdɑsɪˌe] *n.* 全套檔案

我們待會兒再回來討論這些電子郵件，特別是說明你畫了底線的字串。不過你是否已看出以上電子郵件的類型，**電子郵件①、②、⑤是回信，電子郵件③、④是發信。**

電子郵件的常見訊息類型很多，包括提醒、確認、回覆、請求等。寫電子郵件時，你可以在開端以特定的 set-phrases，表明所寫信件的訊息類型，提高收件者處理事情的效率。繼續往下看之前，請瀏覽一下本章的主要學習目標。看完本章，你應該達到的學習目標如下：

☐ 用不同的 set-phrases 表明所寫電子郵件的類型是發信還是回信。
☐ 用 set-phrases 寫不同訊息類型的電子郵件，包括：
  • 提醒
  • 確認
  • 道歉
  • 宣布
  • 回覆壞消息
  • 告知壞消息
☐ 正確使用關鍵的商業名詞—— order 和 opportunity。

好，我們接著進入本章第一節。

回信

　　電子郵件就像書寫形式的對話，遵照著發話－回話的模式。我將在本節中教你收到一封必須回覆的電子郵件時，可以用來回信的常見 set-phrases。現在請做 Task 2.2。記住，做完 Task 後再檢查答案。

Task 2.2

請將下列 set-phrases 分門別類，並填入下表中。

- Thank you for sending me n.p. ...
- Apologies for the delay in getting back to you, but ...
- Thank you for purchasing n.p. ...
- Sorry for the delay in Ving ...
- I apologize for the delay in getting back to you, but ...
- We're sorry to learn (from X) of n.p. ...
- Thank you for your message.
- I'm sorry to learn (from X) about n.p. ...
- Thank you for your email about n.p. ...
- Apologies for the delay.
- Sorry for not getting back to you earlier, but ...
- I am very sorry to learn (from X) that + n. clause ...
- I was sorry to hear about n.p. ...
- Sorry for the delay in responding to ...
- Thanks for your reply.
- I was sorry to hear that + n. clause ...
- Thank you for your query about n.p. ...

道謝 Thanking

回應壞消息 Responding to Bad News

對延遲致歉 Apologizing for Delay

Task 2.2 參考答案

請研讀下面的語庫 2.1，並檢查答案。

道謝 Thanking

- Thank you for purchasing n.p. ...
- Thank you for your message.
- Thanks for your reply.
- Thank you for your email about n.p. ...
- Thank you for your query about n.p. ...
- Thank you for sending me n.p. ...

回應壞消息 Responding to Bad News

- We're sorry to learn (from X) of n.p. ...
- I'm sorry to learn (from X) about n.p. ...
- I am very sorry to learn (from X) that + n. clause ...
- I was sorry to hear about n.p. ...
- I was sorry to hear that + n. clause ...

對延遲致歉 Apologizing for Delay

- Apologies for the delay in getting back to you, but ...
- Sorry for the delay in Ving ...
- I apologize for the delay in getting back to you, but ...
- Apologies for the delay.
- Sorry for not getting back to you earlier, but ...
- Sorry for the delay in responding to ...

💡 語庫小叮嚀

- Thanking set-phrases 是回信時的基本用語，不須用到其他種類的 set-phrases 時，便能派上用場。
- Sorry 一字可以用作拖延事情時的道歉用語，或者聽到壞消息時的反應。注意，responding to bad news 的電子郵件類型中，that 後面是接 n. clause，n.p. 前面則是接介系詞。
- 注意，apologizing for delay 這一類的電子郵件中，I apologize ... 中的 apologize 是動詞，Apologies for ... 中的 apologies 則是複數名詞。

挑一些陌生、很難或新的 set-phrases，利用幾分鐘仔細抄下來，以加深印象。抄完後，接下來請做 Task 2.3。

📱 **Task 2.3**

更正下列句子中的錯誤，然後試著在完整的 set-phrases 底下畫線。見範例①。

① Apologies for the delay getting back you but I have been away.

→ *Apologies for the delay in getting back to you, but I have been away.*

② Apologize the delay.

→ _____

③ I am very sorry to learn that the report.

→ _____

④ I apologies for the delay in getting back to you, but my computer was broken.

→ _____

⑤ I was sorry to hear about you didn't get the project approved.

→ _____

⑥ I was sorry to hear that the project cancellation.

→ _____

⑦ I'm sorry learn about your accident.

→ _____

⑧ Sorry for delay responding to you message.

→ _____

⑨ Sorry for not back to you earlier, but I was ill for a week.

→ _____

⑩ Sorry the delay in respond to your email, but I have been very busy.

→ _____

⑪ Thank you for your email about we are having the meeting.

→ _____

⑫ Thank you to purchase the Excel Inkjet Printer.

→ _____

⑬ Thanks for you reply.

→ _____

⑭ We're sorry to learn from Mary of you will be leaving.

→ _____

⑮ Thank you for sending to me the figures.

→ _____

📑 **Task 2.3** 答案

② Apologies for the delay.

③ I am very sorry to learn about the report.

④ I apologize for the delay in getting back to you, but my computer was broken.

⑤ I was sorry to hear that you didn't get the project approved.

⑥ I was sorry to hear about the project cancellation.

⑦ I'm sorry to learn about your accident.

⑧ Sorry for the delay in responding to your message.

⑨ Sorry for not getting back to you earlier, but I was ill for a week.

⑩ Sorry for the delay in responding to your email, but I have been very busy.

⑪ Thank you for your email about the meeting.

⑫ Thank you for purchasing the Excel Inkjet Printer.

⑬ Thanks for your reply.

⑭ We're sorry to learn from Mary that you will be leaving.

⑮ Thank you for sending me the figures.

　　希望你能將所有錯誤一網打盡，正確地改正了所有錯誤。記住，電子郵件中的錯誤多半出在 set-phrases 的細節之中。如果挑不出錯誤，或者不知從何更正起，可以回到前言第 32 頁，重新複習 Task 9。現在建議你做一個延伸 Task：拿出幾封以前所寫的電子郵件，看一看裡面有沒有這些 set-phrases，檢查用法是否正確。如果有誤，該如何改正？接下來請做下面的 Task。

Task 2.4

請從語庫 2.1 挑出適當的 set-phrases 填入空格中，以完成下列電子郵件。

①

Dear Tracy,

_____ but we've been very busy here with the merger. I have talked to my supervisor about your question and he thinks a better time for the seminar might be next month. I hope this is OK.

②

Dear Tony,

_____ your company is downsizing and that you are probably going to be laid off. Please send me your resume and let me see if I can help you get a job in our company.

③

Dear Lucy,

_____ me the figures. I'll look through them
later and get back to you.

④

Dear Jonathan,

_____ Yes, I will be attending the conference in
Singapore on the 17th and hope to see you there.

⑤

Dear Tom,

_____ the problem with the machinery you
bought from us last month. I can assure you we will do everything we
can to repair it for you as soon as possible.

⑥

Dear Sam,

_____ I have attached the file you requested.
Talk to you later.

⑦

Dear Winston,

_____ but we have had terrible problems with
the new software and have not been able to access any of our data.

① Apologies for the delay in getting back to you,

　Sorry for the delay in replying,

　I apologize for the delay in getting back to you,

　Sorry for not getting back to you earlier,

② I was sorry to hear that

　We're sorry to learn from Mary that

　I am very sorry to learn that

③ Thank you for sending

④ Thank you for your message.

　Thanks for your reply.

　Thank you for your email about the conference.

⑤ We're sorry to learn of

　I'm sorry to learn from Charlie about

　I was sorry to hear about

⑥ Thank you for your message.

　Thanks for your reply.

　Thank you for your email about the project.

　Thank you for your query about the project.

⑦ Apologies for the delay in getting back to you,

　Sorry for the delay in responding,

　I apologize for the delay in getting back to you,

　Sorry for the delay in responding to your query,

這些 set-phrases 當然也可以合併，尤其是 thanking set-phrase 後面可以接 apologizing set-phrase。範例見 Task 2.1 的郵件。

　　目前爲止我們已經做過一些檢視 set-phrases 細節的 Task，也練習過用 set-phrases 寫電子郵件，現在請在接下來的延伸寫作題中寫出四封電子郵件，兩封爲回信、兩封爲發信。若你的時間有限，則兩類電子郵件各寫一

封，以後再回來寫另外兩封或作修改。請耐著性子，不要跳過 Task 直接看答案。工欲善其事，必先利其器，多練習才能提升寫作能力。

延伸寫作題 2.1A

請閱讀下列電子郵件，然後看摘要。看完摘要後，請寫一封回信。

①

> Dear Rudi,
>
> After discussion and cost benefit analysis, we have decided to use your company for all our computing needs for the next three years. Our lawyers are drawing up a contract, which you can expect to receive in the next few days.
>
> I look forward to doing business with you in the future.
>
> Best regards,
> Trudi

②

> Dear Kirsty,
>
> Where are you? I have written three emails to you and received no reply. I have an urgent enquiry regarding the ABC project. Are we going ahead with this or not? And if so, who is going to lead it?
>
> Please contact me as soon as you can.
>
> Best wishes,
> Ming

Vocabulary

draw up 起草

💡 Email 摘要

1. 楚蒂寫信通知魯迪，楚蒂的公司將把未來一切的運算需求交給魯迪的公司來打點。她的律師正在草擬合約，擬好就會寄給魯迪。

2. 明寫信給克絲媞。他很擔心，因為他完全沒有收到她回覆問題的回信。他想要知道，ABC 的案子要不要做，要做的話又是由誰來當主持人。

 現在請以下面的參考答案核對你所寫的回信。閱讀時，不妨練習將目前學過的 set-phrases 畫底線，增加你對它們的記憶。

📑 延伸寫作題 **2.1A** 參考答案

①

Dear Trudi,

Thank you for purchasing our service. We are happy that you have chosen us for your computing needs. I look forward to receiving the contract. In future, if you have any questions or concerns regarding your account, please do not hesitate to contact me.

Sincerely,
Rudi

---

②

Dear Ming,

I apologize for the delay in getting back to you, but I've been away having a baby. I am back in the office now so I will give your query my first priority. You can expect my answer by the end of the week.

Best wishes,
Kirsty

請閱讀下列摘要，然後看翻譯。看完翻譯後，請寫兩封電子郵件。

①

Martin is writing to Marilyn to offer to help her look for a job in his company. Marilyn recently informed Martin that her company was downsizing and that she might be made redundant.

翻譯

馬丁向瑪莉蓮提議，幫她在自己公司找工作。最近瑪莉蓮告訴馬丁，她的公司要縮編，她可能會被資遣。

②

Eva is writing to Montgomery to try to calm him down. Montgomery has been writing angry emails to Eva recently because she did not reply to an urgent request he sent her about six weeks ago. Eva explains that while she was on a two month job swap in New York, no one forwarded her emails to her. She seems to remember telling Montgomery about her job swap before she left for New York. She promises to answer Montgomery's queries when she settles back into her Taiwan office.

翻譯

伊娃寫信安撫蒙戈馬利的情緒。近來蒙戈馬利多次寫信指責伊娃，六個星期前他寫電子郵件給她有急事相求，卻遲遲沒有回音。伊娃解釋道，因為她過去兩個月到紐約交換工作，卻沒有人將電子郵件轉寄給她。她隱約記得去紐約前告訴過蒙戈馬利交換工作的事。她保證回到台灣公司時，一定會回覆蒙戈馬利的問題。

Vocabulary

swap [swɑp] *n.*【口語】交換

①

Dear Marilyn,

Sorry to hear that your company is downsizing. I sincerely hope that you will not be laid off. Please send me your resume and I'll see if there are any openings here. Let me know what else I can do for you.

Best wishes,
Martin

②

Dear Montgomery,

Thanks for your emails, which I did receive. I'm sorry about the delay in responding to you, but I have actually been away on a job swap in New York for two months, and no one in my office took the responsibility to forward your emails to me. I also seem to remember telling you about my job swap a few months ago. Perhaps you simply forgot. Anyway, please give me a few days to settle back in here, and I will deal with your request as soon as I can.

Thanks for your patience.

Eva

　　你的答案可能會與參考答案有些出入，但希望至少你的 set-phrases 沒用錯。學完了回信的 set-phrases，接著可以進入發信 set-phrases 的章節了。

Vocabulary

downsize [ˈdaʊnˌsaɪz] *v.* 縮編　　　　**lay off** 解雇

opening [ˈopənɪŋ] *n.* 空缺　　　　**forward** [ˈfɔrwəd] *v.* 轉交；轉寄

發信

　　截至目前為止，你已經學過收到電子郵件後，回信時可使用的 set-phrases，而本節將教你主動發信時可使用的 set-phrases。這些 set-phrases 除了可顯示所寫電子郵件為發信之外，從中也可看出所寫電子郵件的訊息類型。**發信的訊息可分為五大類：交代事情或索取資訊、宣布、提醒、確認和請求確認。**本節的教學重點則為後四者，至於交代事情和索取資訊的電子郵件，我將留待下一章中詳細說明。首先來學一些 set-phrases 吧。請做 Task 2.5。

Task 2.5

請將下列 set-phrases 分門別類，並填入下表中。

- Just to inform you of n.p. ...
- ... are included in the attachment.
- ... has been received.
- Also attached is n.p. ...
- For your information, ...
- Have you remembered to V ...?
- I am pleased to confirm that + n. clause ...
- I have been informed that + n. clause ...
- Just a quick note to confirm n.p. ...
- Please confirm receipt of n.p. ...
- Just to inform you that + n. clause ...
- We have just p.p. ...
- Attached you will find n.p. ...
- We have received n.p. ...
- Just to let you know that + n. clause ...
- Please be informed that + n. clause ...

- Just to remind you to V ...
- Please be notified that + n. clause ...
- Just to confirm that + n. clause ...
- Please confirm if + n. clause ...
- Please confirm our n.p. ...
- Just to remind you that + n. clause ...
- I can confirm that + n. clause ...
- This is to confirm n.p. ...
- Please confirm whether + n. clause ...
- I attach n.p. ...
- Please find the attached n.p. ...
- Just a short note to confirm that + n. clause ...
- Please note that + n. clause ...
- I can confirm we have received ...
- Please remember to V ...
- Thank you for ... which I received today.
- Just a quick note to remind you to V ...
- This is to confirm that + n. clause ...
- We have been asked to V ...
- This is just a quick note to let you know that + n. clause ...
- We hereby confirm that + n. clause ...
- Please confirm that + n. clause ...
- We wish to confirm that + n. clause ...
- This is to confirm that I will be attending n.p. ...
- We wish to confirm the following: n. clause ...
- Please give me a call to confirm that + n. clause ...
- Would you please confirm that + n. clause ...?
- Your request has been successfully processed.

Vocabulary

hereby [ˌhɪrˈbaɪ] *adv.* 由此；藉此   **process** [ˈprɑsɛs] *v.* 處理；加工

告知訊息 Announcing

提醒 Reminding

確認 Making Confirmation

請求確認 Requesting Confirmation

請研讀下面的語庫 2.2，並檢查答案。

Email 必備語庫 2.2

告知訊息 Announcing

- ... are included in the attachment.
- ... has been received.
- Also attached is n.p. ...
- Attached you will find n.p. ...
- For your information, ...
- I attach n.p. ...
- I have been informed that + n. clause ...
- Just to inform you of n.p. ...
- Just to inform you that + n. clause ...
- Just to let you know that + n. clause ...
- Please be informed that + n. clause ...
- Please be notified that + n. clause ...
- Please find the attached n.p. ...
- Please note that + n. clause ...
- Thank you for ... which I received today.
- This is just a quick note to let you know that + n. clause ...
- We have been asked to V ...
- We have just p.p. ...
- We have received n.p. ...

提醒 Reminding

- Have you remembered to V ...?
- Just a quick note to remind you to V ...
- Just to remind you that + n. clause ...
- Just to remind you to V ...
- Please remember to V ...

確認 Making Confirmation

- I am pleased to confirm that + n. clause ...
- I can confirm that + n. clause ...
- I can confirm we have received ...
- Just a quick note to confirm n.p. ...
- Just a short note to confirm that + n. clause ...
- Just to confirm that + n. clause ...
- This is to confirm that I will be attending n.p. ...
- This is to confirm that + n. clause ...
- This is to confirm n.p. ...
- We hereby confirm that + n. clause ...
- We wish to confirm that + n. clause ...
- We wish to confirm the following: n. clause ...
- Your request has been successfully processed.

請求確認 Requesting Confirmation

- Please confirm if + n. clause ...
- Please confirm our n.p. ...
- Please confirm receipt of n.p. ...
- Please confirm that + n. clause ...
- Please confirm whether + n. clause ...
- Please give me a call to confirm that + n. clause ...
- Would you please confirm that + n. clause ...?

語庫小叮嚀

- 研讀各類訊息的 set-phrases 時，看看能否從中歸納出一些模式？舉例：通常 just 後面是加 to。
- 每一類訊息的 set-phrases 中，n.p.、n. clause 和 V 遵循的模式如出一轍，你看得出是什麼模式嗎？但是 confirm 後接的 n.p. 模式稍有不同，又是如何的不同呢？
- 注意，這些訊息類型中，經常出現完成式（have/has + p.p.），我將在下一章中說明原因。

請挑一些陌生、很難或新的 set-phrases，花幾分鐘仔細地抄寫，以加深印象和記憶。完成後，接下來請做 Task 2.6。

Task 2.6

請選擇最適當的選項以完成下列句子，見範例 ①。

___b___ ① Just to inform you of _____.

    a) you can attend the gala dinner on Friday

    b) the new situation in the Southern Sales Region

_____ ② Just to inform you of _____.

    a) our decision to close the factory

    b) you need to complete your report earlier this month

_____ ③ Just to inform you that _____.

    a) we have decided to close down the factory

    b) our decision to close down the factory

_____ ④ Just to remind you that _____.

    a) our meeting tomorrow afternoon

    b) you need to complete your report earlier this month

_____ ⑤ Just to remind you to _____.

    a) contact the client as soon as you get back to the office

    b) appointment tomorrow afternoon

_____ ⑥ Just a quick note to confirm _____.

    a) our meeting tomorrow afternoon

    b) appointment tomorrow afternoon

_____ ⑦ Just a short note to confirm that _____.

    a) our decision to close down the factory

    b) I will be arriving at 8 pm

_____ ⑧ Please confirm that _____.

    a) contact the client as soon as you get back to the office

    b) you have received the file

_____ ⑨ Please confirm our _____.

    a) I will be arriving at 8 pm

    b) appointment tomorrow afternoon

_____ ⑩ Please confirm if _____.

    a) you would like me to book you a hotel

    b) the new situations in the Southern Sales Region

_____ ⑪ Please confirm whether _____.

    a) you can attend the gala dinner on Friday

    b) contact the client as soon as you get back to the office

Task 2.6 答案

② a   ③ a   ④ b   ⑤ a   ⑥ a   ⑦ b   ⑧ b   ⑨ b   ⑩ a   ⑪ a

    希望你的 n. clause 和 n.p. 用對了，這對寫出正確和清楚明瞭的電子郵件十分重要。上面 Task 的目的是引導你的注意力，專注在 n. clause 和 n.p. 的模式上。我們再做一個 Task，這次將注意力集中在 confirming set-phrases 其他重要的細節上。

請更正下列句子中的錯誤。見範例①。

① I am pleased confirm that we can go ahead with the deal.

→ *I am pleased to confirm that we can go ahead with the deal.*

② Just short note to confirm that the project is on schedule.

→ _____

③ Please confirm a receipt of the attached document.

→ _____

④ Please confirm weather you would like someone to pick you up.

→ _____

⑤ We wish confirm that payment has been received.

→ _____

⑥ Please confirm that payment received.

→ _____

⑦ Please give me a phone to confirm that you have received my email.

→ _____

⑧ This to confirm that I will be attending the meeting on the 29th.

→ _____

⑨ This is confirm that the meeting is cancelled.

→ _____

⑩ Please confirm you need more information.

→ _____

⑪ We here confirm that the project is complete according to our agreement.

→ _____

⑫ We wish to confirm following: XYZ, 123 are the new codes for XYS 124.

→ _____

Task 2.7　參考答案

② Just a short note to confirm that the project is on schedule.

③ Please confirm receipt of the attached document.

④ Please confirm whether you would like someone to pick you up.

⑤ We wish to confirm that payment has been received.

⑥ Please confirm that payment has been received.

⑦ Please give me a call to confirm that you have received my email.

⑧ This is to confirm that I will be attending the meeting on the 29th.

⑨ This is to confirm that the meeting is cancelled.

⑩ Please confirm if you need more information.

⑪ We hereby confirm that the project is complete according to our agreement.

⑫ We wish to confirm the following: XYZ, 123 are the new codes for XYS 124.

　　希望你改正了所有的錯誤。好，既然做完了一些針對 set-phrases 細節的 Task，接下來可以練習將 set-phrases 用在電子郵件中了。

請參見語庫 2.2，將適當的 set-phrases 填入下列電子郵件的空格中。

①

Dear Alison,

_____ turn out the lights when you go home tonight. You left them on last night.

②

Dear All,

_____ I will be on leave for the next two weeks. Carrie will take care of your concerns.

③

Dear Margaret,

_____ the project we discussed last week has been given the go-ahead by head office. Can we get together sometime this week to discuss how to proceed?

④

Dear Tom,

_____ your email account is now working.

Vocabulary

on leave 休假中　　　　　　go-ahead [ˈgoəˌhɛd] *n.* 許可；允許

⑤

Dear All,

_____ the new flyers are ready for distribution. Please pick it up on your way to the shop floor.

⑥

Dear Sharon,

_____ our meeting tomorrow at 3 pm. I'm looking forward to seeing you then.

⑦

Dear Lulu,

_____ you are still using the computer system on a regular basis or not.

⑧

Dear Ms. Chu,

_____ we have received your payment.

⑨

Dear Sales Staff,

_____ accurately record all your figures at the end of every day. There have been too many gaps in the records recently and this is causing problems for the finance department.

① Just a quick note to remind you to

Just to remind you that you need to

Just to remind you to

Please remember to

② For your information,

Just to inform you that

Just to let you know that

Please be informed that

Please be notified that

Please note that

This is just a quick note to let you know that

③ I am pleased to confirm that

I can confirm that

I can confirm we have received notification that

Just a short note to confirm that

Just to confirm that

This is to confirm that

Just to inform you that

I have just heard that

This is just a quick note to let you know that

④ I can confirm that

Just a short note to confirm that

Just to confirm that

This is to confirm that

We wish to confirm the following:

⑤ For your information,

Just to inform you that

Just to let you know that

Please be informed that

Please be notified that

Please note that

This is just a quick note to let you know that

⑥ Just a quick note to confirm

This is to confirm that I will be attending

This is to confirm

⑦ Please confirm if

Please confirm whether

⑧ I am pleased to confirm that

I can confirm that

Just a short note to confirm that

Just to confirm that

This is to confirm that

We hereby confirm that

We wish to confirm that

We wish to confirm the following:

⑨ Just a quick note to remind you to

Just to remind you to

Please remember to

　　相信你已經看出，有時答案的選擇很多，但是請記住兩件事：第一，不管是 reminding、announcing 或 confirming 的訊息，所用的 set-phrases 必須符合電子郵件的文意。第二，不論是 n. clause、n.p. 或 V，所用 set-phrases 的結尾必須符合句子。希望你都答對了。

　　現在繼續練習本章的第二個「延伸寫作題」。這次請寫完四封電子郵件，這四封信都很短，因此希望你能將四封郵件都寫完（記得重點在於訊息本身）。此外，在這個 Task 當中，你必須應用本章後半段學到的四種發信電子郵件。

延伸寫作題 2.2

請先閱讀下列摘要，然後看翻譯。看完翻譯後，請寫出四封電子郵件。

①

Write to Mandy informing her of the meeting on the 23rd.

翻譯

寫信通知曼蒂，會議是在 23 日開。

②

Remind Rockie that he needs to ask his client if the client wants to attend the Christmas event. Tell him that there are only three more places available for this event, and that he should hurry his response.

翻譯

提醒洛基，他需要去問客戶的是，客戶要不要參加耶誕節的活動。告訴他說，這場活動只剩下三個位子，所以他應該要趕緊回覆。

③

Write to Ben confirming that you can give him the 10% discount he was asking for. Inform him that you are attaching a revised order sheet with new prices.

翻譯

寫信跟班確認，對於他所要求的一成折扣，你可以給他。通知他說，你會附上改好新價格的訂貨單。

④

Ask Martin to confirm whether he needs a vegetarian meal on his flight to New York next week.

翻譯

請馬丁確認一下，在下星期往紐約的班機上，他需不需要吃素食餐。

①

Dear Mandy,

Just to inform you of the meeting on the 23rd. See you then!

②

Dear Rockie,

Have you remembered to ask the client whether they would like to attend the Christmas event? We only have three more places available! Please let me know ASAP.

③

Dear Ben,

I am pleased to confirm that we can offer you the 10% discount you requested at our last meeting. Please find the attached revised order sheet with pricing.

Let me know if you have any questions.

④

Dear Martin,

Please confirm whether you need a vegetarian meal on your flight to New York next week.

Thank you.

Vocabulary

order sheet 訂貨單

商務電子郵件中的兩個關鍵名詞為 opportunity 和 order。下列的 word partnership 表中，第一欄是最常與商務電子郵件中此兩名詞搭配的動詞，而中間欄是最常搭配的形容詞。發現了嗎？只要利用下表，即可輕鬆又迅速地造出漂亮的句子。請參見表下的例句並練習造出自己的句子。

Unit
2

動詞	形容詞	名詞
be on the look out for	excellent	
come across	golden	
exploit	great	
find	lost	
get	missed	
give	marvelous	**opportunity**
have	rare	
lose	unexpected	
take advantage of	unprecedented	
wait for		
waste		

例 I don't want to wait for another opportunity.
　　我不想等待另一個機會。

例 We are always on the look out for a rare opportunity.
　　我們總在留意一個難得的好機會。

Vocabulary

be on the look out for 留意 　　　　　　**exploit** [ɪkˋsplɔɪt] *v.* 利用

marvelous [ˋmɑrvələs] *adj.* 令人驚歎的；很棒的

unprecedented [ʌnˋprɛsə͵dɛntɪd] *adj.* 無先例的；空前的

動詞		形容詞	名詞
accept	place	back	
authorize	process	bulk	
cancel	receive	initial	
change	send	new	
chase up	ship	regular	
check	take	repeat	**order**
confirm		special	
dispatch		urgent	
fax through		wholesale	
fill		large	
increase		small	

例 Who authorized this bulk order?
　這批大訂單是誰授權的？

例 I'm just faxing through your order now.
　我正在傳真你的訂單。

　　現在你已經學過回信和發信的 set-phrases，不妨拿出幾封以前寫的電子郵件，嘗試放入本章所學到的 set-phrases，看看能否更清楚地表達訊息。你在那些電子郵件中是否使用了本章教的 set-phrases？如果有，這些 set-phrases 是否正確？

　　本章即將結束，但是進入下一章前，請溫習前言中的電子郵件和你畫了底線的字串。你能看出自己進步了多少嗎？再看看本章前言中所設定的學習目標，勾出自認為達成的項目。如果有尚未學會的項目，請過幾天後重新閱讀本章，看看是否更能理解內容。

Vocabulary

authorize [ˈɔθəˌraɪz] *v.* 授權；批准
dispatch [dɪˈspætʃ] *v.* 急派；迅速處理

chase up 追查
bulk [bʌlk] *adj.* 整批的；大量的

Unit 3

表達請求的 Email

　　上一章中，我提過主動發信的電子郵件中，交代事情或索取資料、宣布、提醒、確認和請求確認等訊息類型最為常見。本章的教學重點為請求信（request email）中，交代事情或索取資料時會用到的字串，以及要寫好一封請求信的其他要素，例如：說明交代事情的背景、原因和告知期限。本章一開始，我們先看一封典型的請求信吧。

Task 3.1

請閱讀下列郵件。在交代事情的字串下畫直線，在索取資訊的字串下畫曲線。

Dear Marjorie,

I have just received a call from customer XYZ asking me for information on his account. Since our computer system is broken again, I wonder if you could help me with this.

Could you let me know how many payments the customer made last month, i. e., from August 1 to August 30? Also, can you tell me how many of those payments were over NT$10,000, and how many were under NT$10,000? As I need this information really quickly, would it be possible for you to ask someone to bring a printout of this information over to me ASAP? I really would appreciate it. Thanks for your help.

Best wishes,
Alvin

Vocabulary

printout [ˋprɪntˌaʊt] *n.*【電腦】資料的印出；所印出的資料

💡 Email 摘要

艾文向瑪卓莉索取一名客戶的帳戶資料。艾文的電腦系統當掉,需要瑪卓莉幫忙。艾文要索取該名客戶的繳款次數和款項數目等詳細資料,另希望瑪卓莉列印出該名客戶的帳戶明細後,盡快派人拿給他。

我稍後會回頭討論這封電子郵件,以及你畫了線的字串。現在先看看本章的學習目標。看完本章,你應該達到的學習目標如下:

☐ 寫請求信,交代事情。

☐ 寫請求信,索取資料。

☐ 提供請求的背景資訊。

☐ 說明請求的原因。

☐ 說明請求的處理期限。

☐ 組織請求信的結構,方便收件者閱讀。

Unit
3

交代事情

　　現在來看看交代事情（request for action）時使用的字串吧。首先請看下列語庫，然後做語庫下方的 Task。

Email 必備語庫 3.1

- Can you ...?
- Could you ...?
- Do you think you could ...?
- Do you think you could possibly ...?
- I wonder if you could ...
- I was wondering if you could ...
- Would it be possible for you to ...?

語庫小叮嚀

- 注意哪些 set-phrases 以問號結尾，哪些以句點結尾。
- 在 Can you ...? 和 Could you ...? 的句子中，句首或句尾也可以加上 please，讓語氣聽起來更有禮貌。
- 注意這些 set-phrases 大部分都用 could，而非 can。

Task 3.2

上列語庫中的字串，哪些字串比較客套？哪些語氣較為直接？

Task 3.2 參考答案

希望你可以看出，有些 set-phrases 比較長，有些比較短。記住，一般 set-phrases 的禮貌程度有個粗略的常規，就是**愈長的 set-phrases 愈有禮貌**，**愈短的 set-phrases 愈直接**。

例如，Can you ...? 很短，語氣頗爲直接，但仍然不失禮貌；Do you think you could possibly ...? 這句子長得多，語氣更有禮貌。**寫電子郵件時，選擇 set-phrases 的關鍵在於所交代事情的性質，而非你和收件者之間的關係。**舉例來說，若要請對方幫你寫一份重大報告，由於這非小事而且是很嚴肅的請求，這時應該用較長的 set-phrases。但是如果僅僅是請求收件者寄檔案給你，於對方來說是舉手之勞，用短的 set-phrases 就可以了。

很多人擔心不同的 set-phrases 會有不同的適用對象；有的針對上司，有的針對上司的上司，有的則針對下屬，如果用錯對象，將嚴重冒犯對方。這種擔心是多餘的，就如我在＜前言＞中所說，商界人士最關心的是電子郵件的訊息，而非語氣有禮與否。因此記住，選擇 set-phrases 時，請根據所交代事情的性質，並且清楚傳達訊息才是寫電子郵件時永遠不變的重點。

言歸正傳，提出請求時該如何使用 set-phrases？很容易，只要將 set-phrases 放在請對方處理的事情之前，就可以了。如果你交代的事情是 Bring me the file.，就把 set-phrase 放在那個句子之前，變成：Can you bring me the file?。一點也不難吧！但是容易歸容易，別忘記改變標點符號。現在請做下面的 Task。

Task 3.3

請改寫下列祈使句，使成爲交代事情的請求句，見範例 ①。

① Bring it over.

→ *Can you bring it over?*

② Look at my budget figures.

→ _____

③ Complete this report.

→ _____

④ Tell John that I need the data now.

→ _____

⑤ Give me an answer soon.

→ _____

⑥ Send this document to New York as quickly as possible.

→ _____

⑦ Pick me up at the airport.

→ _____

⑧ Prepare the data I need for my presentation.

→ _____

⑨ Cover for me on Tuesday while I take the day off.

→ _____

現在請核對以下答案。

Task 3.3　參考答案

② Could you look at my budget figures?

③ I was wondering if you could complete this report.

④ Can you tell John that I need the data now?

⑤ Could you give me an answer soon?

⑥ Do you think you could send this document to New York as quickly as possible?

⑦ Would it be possible for you to pick me up at the airport?

⑧ I was wondering if you could prepare the data I need for my presentation.

⑨ I wonder if you could cover for me on Tuesday while I take the day off.

為了加強你對細節的注意，請改正下一個 Task 句子中的錯誤，每句的錯誤可能不只一個。

Task 3.4

請改正句子中的錯誤，見範例 ①。

① Can you bring it over.

→ *Can you bring it over?*

② Can you get back to me as soon as possible please.

→ _____

③ Do you think you could possible ask someone from accounts to contact me about this.

→ _____

④ Do you think you could spare some time to go through this with me.

→ _____

⑤ Would it be possible you provide me with more data on this problem?

→ _____

⑥ I wonder you could send someone to look at the machine and find out what the problem is?

→ _____

⑦ Could you phone him for me, please.

→ _____

⑧ I was wondering you can prepare the PowerPoint presentation for me?

→ _____

現在請核對以下答案。

② Can you get back to me as soon as possible, please?

③ Do you think you could possibly ask someone from accounts to contact me about this?

④ Do you think you could spare some time to go through this with me?

⑤ Would it be possible for you to provide me with more data on this problem?

⑥ I wonder if you could send someone to look at the machine and find out what the problem is.

⑦ Could you phone him for me, please?

⑧ I was wondering if you could prepare the PowerPoint presentation for me.

✅ 索取資料

好，接下來我們要學的是索取資料時使用的字串，這比交代事情時使用的字串稍微複雜一點。請先在紙上寫下五個句子，向同事或客戶索取資訊，待會兒我們再回來討論這些句子。現在請做下面的 Task。

📑 Task 3.5

請研讀下列語庫。交代事情和索取資料的字串有何相似和不同之處？

Email 必備語庫 3.2

- Can you tell me 'wh' ...?
- Do you think you could possibly tell me 'wh' ...?
- Would it be possible for you to let me know 'wh' ...?
- I wonder if you could explain 'wh' ...
- Could you explain 'wh' ...?
- I was wondering if you could let me know 'wh' ...
- Do you think you could let me know 'wh' ...?

你看出差異了嗎？現在請以下面的語庫小叮嚀來核對答案吧。

語庫小叮嚀

- 注意，所有 set-phrases 的開頭和交代事情的 set-phrases 是一樣的。
- 注意，在開頭的 set-phrases 之後可以接三種不同的 chunks：... let me know ...、... tell me ... 和 ... explain ...（不是 ... explain me ...）。
- 注意，所有的 set-phrases 最後都以 wh- 字結尾。所謂的 wh- 字就是：who、why、what、when、where、why、which 和 how。
- 以 can 或 could 開頭的請求句，何時該用句點，何時該用問號？其實規則很簡單，當提問者真正在詢問對方的意願並期望對方給個答案時，用問號；當提問者並不期望對方的回答，而是要求對方接受這個提議時，用句點。

Unit
3

　　如同交代事情的 set-phrases 一樣，索取資料的 set-phrases 用法也很簡單，set-phrases 只要放在索取資料的句子之前準沒錯。換句話說，如果你原本要問對方：What time is the conference starting?，你可以寫成：Could you tell me what time the conference is starting?。但是使用索取資料的字串時，須將五大原則謹記於心：

原則一：主詞和動詞的位置務必記得交換。

　　請看以下範例，你就會明白我的意思。

1. When **is the conference** starting?

   Can you tell me when **the conference is** starting?

2. Which hotel **have you** reserved for me?

   Could you let me know which hotel **you have** reserved for me?

3. Where **is the conference venue**?

   I was wondering if you could explain where **the conference venue is**.

原則二：如果問句的主詞是 **who**，主詞不必和（助）動詞易位。
見以下範例。

1. **Who is** giving the opening speech?
   Do you think you could let me know **who is** giving the opening speech?
2. **Who will** pick me up from the airport?
   Can you tell me **who will** pick me up from the airport?

原則三：如果原始問句包含 **do**、**does** 或 **did** 等字，這些字不要用在請求句中，主詞和動詞也不要易位。

見以下範例。注意問句加上 set-phrases 時動詞的變化。（若 do、does、did 為否定型式，則助動詞置於主詞後面。）

1. Why **does** the total **include** the figures for last quarter?
   Could you tell me why the total **includes** the figures for last quarter?
2. Where **do** you normally **take** clients to dinner?
   I was wondering if you could tell me where you normally **take** clients to dinner.
3. What **did** the board **say** to our proposal?
   Do you think you could tell me what the board **said** to our proposal?

原則四：如果原始問句不是以 **wh-** 字為句首，而是以 **yes-no** 問句開頭，這些原則仍然適用，但必須插入 **whether**。

※ 用 if 也可以，但建議還是用 whether 比較好，因為用 if 的話，可能會產生另一種句意，例如：Can you tell me if you are going to the conference?（如果你要來參加會議，能告訴我嗎？）見以下範例。

1. Are you going to the conference?
   Can you tell me **whether** you are going to the conference?
2. Do you still need this data?
   I was wondering if you could let me know **whether** you still need this data.

3. Did you include last year's figures in the total or not?

Could you explain **whether** you included last year's figures in the total or not?

原則五：最後，別忘了標點符號的重要性**，這個原則我們之前已經討論過了。**

現在快速地複習一下這五大原則，確定你真的都懂了，再繼續做下一個 Task。

1. 記得動詞、主詞要易位。
2. 記得遇到 who 問句時，動詞和主詞的位置維持不變。
3. 記得原始問句中的 do、does 或 did 要省略。
4. 記得遇到 yes-no 問句時要用 whether。
5. 記得標點符號不要用錯。

一下子在腦中塞進這麼多原則，有點吃力吧？或許目前這些原則看來有點複雜，但是關鍵就在勤加練習。你也可以多花點時間研讀 Task 3.6 和 3.7 的參考答案。現在請做下一個 Task，將剛剛學過的原則融會貫通。

Task 3.6

請將下列問句改寫成索取資訊的請求句，見範例①。

① When is the meeting starting?

→ *Can you tell me when the meeting is starting?*

② Why are these figures higher than we expected?

→ _____

③ Who is responsible for new customer accounts?

→ _____

④ Which set of figures is the correct one?

→ _____

⑤ Are you going to New York on Tuesday?

→ _____

⑥ Do you have any information on this?

→ _____

⑦ How have you calculated the results from last quarter?

→ _____

⑧ Who gave you permission to park in my space?

→ _____

⑨ When will we be able to see some results in your area?

→ _____

Task 3.6 參考答案

② I was wondering if you could explain why these figures are higher than we expected.

③ Could you let me know who is responsible for new customer accounts?

④ Do you think you could let me know which set of figures is the correct one?

⑤ Can you let me know whether you are going to New York on Tuesday?

⑥ Would it be possible for you to let me know whether you have any information on this?

⑦ I wonder if you could explain how you have calculated the results from last quarter.

⑧ Do you think you could possibly tell me who gave you permission to park in my space?

⑨ Can you let me know when we will be able to see some results in your area?

若你選擇放在原始問句句首的 set-phrases 與參考答案不同，沒關係，記住前面提到的五大原則才是最重要的。

Task 3.7

請改正句子中的錯誤，見範例 ①。每個句子可能不只一個錯誤。

① Can you tell me why are the figures higher than they were last quarter.

→ *Can you tell me why the figures are higher than they were last quarter?*

② Do you think you could possibly tell me what is the password?

→ _____

③ Would it be possible for you let me know when are you coming back to the office?

→ _____

④ I wonder if you could explain who responsible is for all these different jobs?

→ _____

⑤ Could you let me know what does this number mean?

→ _____

⑥ I was wondering if you could let me know you will be bringing your family to the conference?

→ _____

⑦ Do you think you could let me know you need more data.

→ _____

② Do you think you could possibly tell me what the password is?

③ Would it be possible for you to let me know when you are coming back to the office?

④ I wonder if you could explain who is responsible for all these different jobs.

⑤ Could you let me know what this number means?

⑥ I was wondering if you could let me know whether you will be bringing your family to the conference.

⑦ Do you think you could let me know whether you need more data?

　　希望你大部分都答對了。上面句子中所有的錯誤，均與前文所述的五大原則脫離不了關係。如果一個錯誤都找不著，或者不確定爲何有錯，請回頭重新研讀那五大原則，直到了解透徹爲止。現在請拿出之前寫下的五個索取資訊的句子（沒忘記吧？），利用學過的 set-phrases 和五大原則，將其改寫成請求句，當作額外的練習。請現在就動筆吧！

　　好，我們已經學過 set-phrases 的細節和寫法，現在可以嘗試將交代事情和索取資訊的請求句用在電子郵件中了。請做 Task 3.8。

📖 Task 3.8

請在下列電子郵件的空格中，填入適當的交代事情和索取資訊的 set-phrases。

①

Dear Macey,

Thanks for your monthly report, which I received a few days ago. I have read it carefully, but there are a few things I am not clear about, and which I hope you can explain.

_____Ⓐ_____ the sales figures for Q2 are so much lower than the target figures? You do not really make this clear.

Also, _____ B _____ in more detail why you have included the costs for the XYZ project in Q2.

Also, _____ C _____ save the document as a .rtf file and email it to me again. One more thing, _____ D _____ have it translated into Chinese by the end of next week? That would be great. Thanks.

②

Dear Carol,

Thanks for your email regarding your request for professional English training. Generally speaking I think it is a good idea and one that would help to improve the company workflow.

However, _____ E _____ supply me with more information so that I can make a proper decision about it? _____ F _____ you think you need the training and what the topic and content of the training will be? _____ G _____ how long the training is going to last and where the training will be held? _____ H _____ how much we are expected to pay for it and something about the company who is providing the training.

I look forward to hearing from you.

③

Dear Mike,

Thanks for finding us a new supplier at such short notice. I have looked through their proposal and there are still a few points I would like cleared up before I make a decision.

Vocabulary

workflow [ˈwɜːkˌfloʊ] *n.* 工作流程　　　　**clear up** 解決；澄清

_____⟨I⟩_____ their other customers are: I want to make sure we are not using the same supplier as our competitors. Also, _____⟨J⟩_____ discounts they offer for orders over 1,000 units? _____⟨K⟩_____ ask them about their payment terms and conditions.

_____⟨L⟩_____ put all this into a report and let me have it by Monday. Thanks.

Task 3.8　參考答案

①

Ⓐ Can you tell me why

Do you think you could possibly tell me why

Would it be possible for you to let me know why

Could you explain why

Do you think you could let me know why

Ⓑ I wonder if you could explain

I was wondering if you could let me know

Ⓒ I wonder if you could

I was wondering if you could

Ⓓ Can you

Do you think you could possibly

Do you think you could

Would it be possible for you to

Could you

②

Ⓔ Can you

Do you think you could possibly

Do you think you could

Would it be possible for you to

Could you

Ⓕ Can you tell me why

Do you think you could possibly tell me why

Would it be possible for you to let me know why

Could you explain why

Do you think you could let me know why

Ⓖ Can you tell me

Do you think you could possibly tell me

Would it be possible for you to let me know

Do you think you could let me know

Ⓗ I wonder if you could explain

I was wondering if you could let me know

---

③

Ⓘ I wonder if you could explain who

I was wondering if you could let me know who

Ⓙ can you tell me what

do you think you could possibly tell me what

would it be possible for you to let me know what

could you explain what

do you think you could let me know what

Ⓚ I wonder if you could

I was wondering if you could

Ⓛ Can you

Do you think you could possibly

Do you think you could

Would it be possible for you to

Could you

上面的 set-phrases 每一個都可以用，重點是請注意索取資訊和交代事情 set-phrases 的差異、標點符號是否正確，以及沒用錯 wh- 字。

現在我們要做一個延伸寫作題，結束本節的學習。

延伸寫作題 3.1

請閱讀下列摘要，然後看翻譯。看完翻譯後，請寫一封電子郵件。

> Mandy is writing to Robert to thank him for the proposal he has finished. Mandy thinks the proposal is good, but she wants certain questions answered before she hands the proposal up to her boss for approval. She wants more information on:
>
> 1. What are the stages of the project and how are these stages to be implemented?
> 2. What are Robert's figures for growth estimates for next year?
> 3. Who should lead the project from their end?
>
> Mandy wants to make a good impression on her boss, so she would like 5 copies of the proposal, professionally bound in nice looking covers, and she wants translations of the proposal in Chinese, Spanish and Korean.

翻譯

曼蒂寫信給羅勃，以感謝他完成了草案。曼蒂覺得草案不錯，但是在呈給老闆審核前，她有某些疑問想要獲得解答。她想要的更多資訊有：

1. 案子有哪些階段，以及這些階段要如何實施？
2. 羅勃對明年的成長預估數字為何？
3. 對方應該由誰來主持案子？

曼蒂希望讓老闆留下好印象，因此草案需要五份，用精美的封面裝訂起來，並且提供中文、西班牙文和韓文翻譯。

Dear Robert,

Thank you for sending me your proposal so quickly. I have read through it carefully, and before I pass it on to my boss, I just need you to clarify certain things which are not quite clear enough in the proposal yet.

Could you let me know exactly what the stages of the project are and how these stages are to be implemented? Also, I was wondering if you could explain in more detail what the growth estimates for the next year are? Would it also be possible for you to tell me who in your view is suitable to lead the project from your end?

I would like to show your proposal to the management team, so could you please make 5 copies. Do you think you could possibly have the copies bound professionally in some nice covers, to make a good impression? And would it also be possible for you to have the proposal translated into Chinese, Spanish and Korean? I'll need the copies and the translations by the end of next week.

Many thanks for your hard work.

Mandy

說明背景和理由

在請求信中，寄件者往往會在提出請求的同時，簡短地說明 background situation（背景狀況），也可能會提及請求的 reason（原因）。也就是說，寄件者會解釋請對方處理事情或提供資訊的理由。這些資訊有助於對方了解請求的重要性，進而提供必要的協助。本節的教學內容你可能大部分已經學過，但是或許對概念仍然似懂非懂。希望讀完本節後，你會對背景和理由的寫法有更加透徹的了解。現在請做 Task 3.9。

Task 3.9

請閱讀以下郵件，在說明背景的字串下畫直線，在說明理由的字串下畫曲線。

①

Dear Mark,

I have recently received a complaint from customer XYZ about your manner. Because I have received similar complaints from this customer before in connection with other members of our sales team, I am trying to find out what is going on. I wonder if you could let me know whether you have ever tried to offer the customer a bribe, or whether you know of any of our sales team who have tried to do so. Could you also let me know what happened in your last meeting with the customer by the end of today?

Thanks for your help.

Chris

Vocabulary

bribe [braɪb] *n.* 賄賂

Email 摘要

克里斯告訴馬克，一名客戶對馬克提出投訴，說他可能曾經試圖賄賂。但是克里斯也經常收到這名客戶對其他業務員的投訴，因此他將展開內部調查。克里斯問馬克，是否曾經賄賂該名客戶或是其他業務曾如此做過，並詢問上次馬克和該名客戶開會的實際情形。

②

Dear Oliver,

I have recently received a request from a client for more information about our marketing strategy for their product. Because I am going to HK this afternoon, I do not have time to handle their request. Could you please contact the client – details below – to find out what they want to know? Would you also please call me later tonight to let me know what they want?

Thanks,

Emily

Email 摘要

艾蜜莉請奧立弗幫忙。她的客戶來信，向她索取產品的市場策略相關資訊，但她正趕著去香港，只好請奧立弗協助處理該客戶的請求。艾蜜莉請奧立弗晚上打電話給她，告訴她最新情況。

我們等一下再回來討論這些電子郵件。

希望你看得出來，這兩封電子郵件第一句中的 receive 都是現在完成式（have + p.p.），這一句便是在說明背景。下面的語庫中列出說明背景時一些最常用的動詞。通常這些動詞要以現在完成式表示。現在請看語庫，然後做下面的 Task。做 Task 時，別忘了先遮住答案喔！

Email 必備語庫 3.3

achieve	have achieved	develop	
ask sb. to V		find	have found
become		finish	
change		send sth. to sb.	
complete		happen	
receive		improve	
speak to sb.		read	
write to sb.		look through	
decide to V			

Task 3.10

請將語庫中的動詞改成現在完成式（have + p.p.）。

Task 3.10　參考答案

have/has achieved	have/has developed
have/has asked sb. to V	have/has found
have/has become	have/has finished
have/has changed	have/has sent sth. to sb.
have/has completed	have/has happened
have/has received	have/has improved
have/has spoken to sb.	have/has read
have/has written to sb.	have/has looked through
have/has decided to V	

很簡單吧！以前你或許已經學到很多關於 present perfect tense（現在完成式）的用法，但可能概念還是有些模糊。現在我要告訴你幾個簡單的要點，下次寫電子郵件時便知道如何使用這種時態了。

1. 說明請求的背景時，這些動詞必須用現在完成式。
2. 所有上述動詞一定要用現在完成式。
3. 記得要根據主詞是單數或複數，使用 has 或 have。

我還得解釋一下另一個要點，因為很多人犯錯便是栽在這個地方：這個要點和 past perfect tense（過去完成式）或 had + p.p. 有關。我知道以前你在學校時代學過這種時態，可能覺得又難又複雜。現在告訴你一個好消息：商業書信中，不須使用這種時態。所以別再為此傷神，把它拋到腦後吧！

進入下一個 Task 之前，現在正是回顧本章所有電子郵件的時候，注意其中說明背景的句子時態，你能找出語庫中的動詞嗎？有幾個？哪些出現的頻率最高？請現在就檢視本章的所有電子郵件。完成後請繼續做下一個 Task。

Task 3.11

請利用語庫 3.3 中的動詞，將其改成正確時態，填入以下空格中。

① It _____ necessary to upgrade our records.

② We _____ our billing system.

③ We _____ our registration procedure and are now ready for business.

④ Joe _____ me to write to all the department heads to provide information on this issue.

⑤ The company _____ to close the department.

⑥ The vendor _____ to me asking for more information.

⑦ I _____ a virus in our system.

⑧ A lot of things _____ at our end recently.

① has become
② have changed / completed / improved / finished
③ have changed / completed / improved / finished
④ has asked
⑤ has decided
⑥ has written
⑦ have found
⑧ have happened

看得出來嗎？有些句子的空格可以填入的動詞不只一個，但最重要的是你在 have/has 的部分寫對了。

◎ 說明理由

好，現在我們來學理由（reason）的寫法。請回到本節前面的電子郵件，看看你畫了底線的說明理由的字串。這兩封電子郵件說明理由的字串有何共同之處？相信你看出來了，這兩封電子郵件的理由字串都用了 because 一字。表示理由的用字很多，because 不過是其中一個，現在請看看語庫 3.4 中其他表示理由的用字。

Email 必備語庫 3.4

as + n. clause since + n. clause	because + n. clause so + n. clause	because of n. p.

相信你以前學過，也常常用到這些字，因此對這些字都不陌生，但是在此提醒你三個很重要的規則：

1. Because 後面是接名詞子句，但 because of 後面則要接名詞片語。
2. 如果在句子中用了 because，萬萬不可同時又用 so。一個句子中只能兩者擇其一，絕不能兩者兼用。其他表示理由的用字一樣得遵守這個規則。記住，每個句子只能用一個表示理由的用字。
3. 表示理由的用字那麼多，但是記住，一個句子必須有兩個子句。

　　現在請做 Task 3.12，檢驗自己是否已經確實掌握這些規則。

Task 3.12

請改正並重寫下列句子。句子中的錯誤是違反了上述哪一個規則？將規則編號寫在題號前的空格上，並在題目下方寫出正確的句子。見範例①。

　2　① Because I am going away, so I am asking Sandy to take over.

　→ *I am going away, so I am asking Sandy to take over.*

_____ ② Because of there is increased competition in the market, we have decided to reduce the price of the product.

→ _____

_____ ③ As we do not have enough data, so we cannot answer your queries at this time.

→ _____

_____ ④ Because increased competition, we are withdrawing from the market.

→ _____

Vocabulary

take over 接管；繼任　　　　　　**withdraw** [wɪðˈdrɔ] v. 撤退；抽回

⑤ Since there is still so much to do, so we are delaying the project
completion date.

→ _____

⑥ Because we are increasing the size of our order.

→ _____

⑦ Because of we have not received payment, we are holding on to
your order.

→ _____

⑧ Since we do not want to have extra costs.

→ _____

⑨ Because cheaper distribution costs, we should lower the price of
the product.

→ _____

Task 3.12 參考答案

1 ② Because there is increased competition in the market, we have decided to
reduce the price of the product.

2 ③ As we do not have enough data, we cannot answer your queries at this time.

1 ④ Because of increased competition, we are withdrawing from the market.

2 ⑤ Since there is still so much to do, we are delaying the project completion
date.

3 ⑥ Because we are increasing the size of our order, we hope you can give us
a discount.

1 ⑦ Because we have not received payment, we are holding on to your order.

3 ⑧ We are reducing our office floor space since we do not want to have extra
costs.

1 ⑨ Because of cheaper distribution costs, we should lower the price of the
product.

希望你能挑出所有的錯誤，也能看出各題是違反了哪一個規則。現在我們來用這些字串造句吧。接下來請做 Task 3.13。

Task 3.13

利用語庫 3.4 的字串合併下表中左右欄的句子，並將完整句子寫下。見範例 ① 。

1	I wonder if you could help me with this.	a	We are cancelling all outstanding projects.
2	My A.E. will deal with all my correspondence.	b	We are going to have to ask you to resign.
3	I hope you can explain in more detail.	c	We need you to pay your bill as soon as possible.
4	the merger	d	Our computer system is broken again.
5	I need your help with some of the data.	e	There are a few things I am not clear about.
6	We have a cashflow problem.	f	I am going on holiday for a week.
7	the SARS situation	g	We need to complete the project soon.
8	There are problems with the data.	h	We are going to split the company workforce into two separate teams.
9	You have revealed company secrets.	i	All the results need to be readjusted.

Vocabulary

correspondence [ˌkɔrəˈspɑndəns] *n.* 通信；信件往返
cashflow [ˈkæʃˌflo] *n.* 現金流轉

d ① *Because our computer system is broken again, I wonder if you could help me with this.*

②

③

④

⑤

⑥

⑦

⑧

⑨

__f__   ② Since I am going on holiday for a week, my A.E. will deal with all my correspondence.

__e__   ③ As there are a few things I am not clear about, I hope you can explain in more detail.

__a__   ④ Because of the merger we are cancelling all outstanding projects.

__g__   ⑤ Since we need to complete the project soon, I need your help with some of the data.

__c__   ⑥ We have a cashflow problem, so we need you to pay your bill as soon as possible.

__h__   ⑦ Because of the SARS situation, we are going to split the company workforce into two separate teams.

__i__   ⑧ Since there are problems with the data, all the results need to be readjusted.

__b__   ⑨ As you have revealed company secrets, we are going to have to ask you to resign.

> 從參考答案不難看出,每個合併句可以用的 reason 字串很多,希望現在你已經掌握 reason 字串的用法。

再做一個 Task。這次你只需要看看一封請求信的格式,不難吧?

Task 3.14

請回去看看本章的請求信(request email),然後從下列選出最適合請求信的郵件結構(structure)。

Structure 1: Request for action

Reason

Request for information

Background

Unit
3

Structure 2: Background/Reason

Request for action/Request for information

Structure 3: Request for information

Background

Request for action

Reason

Task 3.14 答案

希望你從本章先前的郵件格式中，可以注意到 **Structure 2** 是最適合這類電子郵件的格式。現在請做 Task 3.15，寫 Task 時別忘了先遮住答案。

Task 3.15

請重組下列電子郵件中的句子，按照先後順序，將各句子的編號排列出來。

Dear Emily,

Ⓐ Please could you let me know what the engineer says, and how much the repair is going to cost by the end of this week? Ⓑ If necessary, could you also change the paper. Ⓒ The photocopier has broken down again. Ⓓ I was wondering if you could call the supplier and ask them to come and check the machine and the paper. Ⓔ We have recently changed our photocopying paper to a much cheaper brand, so I think it might have something to do with the new paper.

Thanks for your help.

Anita

正確順序為：_____

Ⓒ   Ⓔ   Ⓓ   Ⓑ   Ⓐ

延伸寫作題 3.2

請閱讀下列摘要,然後看翻譯。看完翻譯後,請寫一封電子郵件。

> You work for a food company. Tests show that one of your new products contains oil which is not safe to eat. Your boss wants to stop manufacturing the product and recall all the orders.
> Write an email to your overseas customer explaining the situation and asking them to return the units they bought last month. Ask them also for the contact details of any of their customers.

翻譯

你任職於一家食品公司。檢驗顯示,你們有一項新產品所含的油食用起來並不安全。你的老闆想要停止製造這項產品,並回收所有的訂貨。

寫封電子郵件向海外的顧客解釋這個情況,並請他們將上個月所進的貨退回來,同時向他們打聽各家顧客的聯絡詳情。

　　閱讀參考答案時,請試著找出本章學過的字串並畫底線。

延伸寫作題 3.2　參考答案

> Dear Tony,
>
> We have recently been conducting tests on the Onion Pancake product (number XYZ123) and these tests have shown that the product may contain unedible oil. As we do not want to take a risk with consumers'

Vocabulary

unedible [ʌnˈɛdəbl] *adj.* 不能食用的

safety or health, we need to carry out a product recall. For this reason, do you think you could return as soon as possible any stock of this item you have left unsold? I was also wondering if you could let me know the names and contact details of any of your customers who have bought this product within the last six months, so that I can contact them about this operation.

Please be assured that all costs of this product recall will be met by us. Your urgent cooperation in this matter is appreciated.

Vocabulary

carry out 實行；執行
return [rɪ'tɜn] *v.* 歸還；退回
meet [mit] *v.* 償還；支付

recall [rɪ'kɔl] *v./n.* 收回；撤消
stock [stɑk] *n.* 庫存；存貨

寫請求信時，如果註明交代事情或索取資料的 due date（期限），讓對方知道請求的緊急程度，可加速處理事情的效率。Due date 的字串用來非常簡單，下列表格提供一些常見寫法。現在請做 Task 3.16。

Task 3.16

請將這些表達時間的 chunks 和用字分門別類，寫在下方表格中。

- ... a few days ago ...
- ... afterwards ...
- ... as soon as possible ...
- ... at once ...
- ... at the end of last quarter ...
- ... at the end of last year ...
- ... before COB today ...
- ... by the 15th of ...
- ... by the end of ...
- ... by the end of next quarter ...
- ... by the end of this quarter ...
- ... by the end of this week ...
- ... immediately ...
- ... in a few days ...
- ... in advance ...
- ... in time for ...
- ... last week ...
- ... later ...
- ... next week ...
- ... no earlier than ...
- ... no later than ...
- ... of the year ...
- ... soon ...
- ... the day after tomorrow ...
- ... the day before yesterday ...
- ... within the next few days ...
- ... yesterday ...
- ... ASAP ...

Vocabulary

COB 營業時間結束（close of business 的縮寫）

Future Time

Past Time	Past or Future Time

Future Time	
◆ ... ASAP ...	◆ ... no earlier than ...
◆ ... as soon as possible ...	◆ ... in advance ...
◆ ... in a few days ...	◆ ... within the next few days ...
◆ ... next week ...	◆ ... at once ...
◆ ... by the end of this week ...	◆ ... later ...
◆ ... by the end of next quarter ...	◆ ... the day after tomorrow ...
◆ ... by the end of this quarter ...	◆ ... before COB today ...
◆ ... no later than ...	

Past Time	Past or Future Time
◆ ... a few days ago ...	◆ ... by the end of ...
◆ ... last week ...	◆ ... of the year ...
◆ ... yesterday ...	◆ ... in advance ...
◆ ... at the end of last quarter ...	◆ ... in time for ...
◆ ... at the end of last year ...	◆ ... soon ...
◆ ... the day before yesterday ...	◆ ... immediately ...
	◆ ... afterwards ...
	◆ ... by the 15th of ...

　　未來時間的 chunks（future time chunks）就是用來指示期限的字串。請回頭看本章的電子郵件，找出說明期限的時間字串並畫底線，同時留意這些字串的用法。請現在就練習。

　　在本章結束前，請再回頭看看在 Task 3.9 中畫了底線的字串，看看自己進步了多少。請現在就照做。另外，建議拿出幾封你以前寫的電子郵件，找出說明背景和理由的字串，尤其是 Email 必備語庫 3.3 中的動詞。看看自己的用法是否正確？如果有錯，該如何改正？最後，結束本章的學習之前，請回到本章前言中所設定的學習目標，看看自己是否達成了所有的項目。

Unit 4
安排會面的 Email

　　商務人士寫電子郵件，很多時候是為了安排或確認會議／會面。本章中，我將教你建議會議時間和地點、拒絕或接受對方建議，以及描述和更改會議時間、地點的寫法；同時指出一般人對英文未來時式用法的常見錯誤，尤其是 will 的用法。讀本章之前，請找出幾封最近寫過的類似電子郵件，在讀完本章後可加以比較，以檢視學習成果。我們這就來看看幾封典型安排會面的電子郵件吧。現在請做 Task 4.1。依照慣例，請先閱讀英文電子郵件，大致看懂之後再看中文摘要。

Task 4.1

請閱讀下列電子郵件，了解情形，然後找出安排和確認會面的 set-phrases，並畫底線。

①

Dear Mary,

We need to arrange a time to meet and discuss the new project. Would Thursday at 3 pm be OK for you? I'm on leave for two days, but I'll be back on Thursday morning.

Joy

②

Dear Joy,

I'm afraid I'm going to be out of town on Thursday. How about the following Monday at 10 am?

Mary

③

Dear Mary,

I'm afraid I'm tied up on Monday at 10, but I'll be free at 12. Why don't we do lunch?

Joy

④

Dear Joy,

Monday lunchtime is fine. I'll come to your office at noon. See you then.

Mary

💡 Email 摘要

喬伊向瑪莉建議星期四下午三點開會討論新案子。瑪莉回拒了提議，因為她那時出差，人不在城裡，她建議下週一早上十點開會。下週一喬伊有空，但是十點有事，因此建議在當天開午餐會議。瑪莉接受了建議。

　　我們等一下會回來討論這些電子郵件。現在請先看一看本章的學習目標。讀完本章，你應該達到的學習目標如下：

☐ 用 set-phrases 提出建議。
☐ 用 set-phrases 接受建議。
☐ 用 set-phrases 拒絕建議。
☐ 用 set-phrases 描述會面時間、地點。
☐ 用 set-phrases 更改會面時間、地點。
☐ 清楚了解這類電子郵件中，be going to 和 will 的用法。

　　現在我們就進入本章第一節，學習安排會面時間、地點的寫法吧！

安排會面

通常安排會面的時候，都是由一人建議會面時間、日期和地點，如果對方不方便，便回信拒絕建議，再提出另外的提議。這樣的過程不斷重複，直到找出雙方都能接受的提議，會面一事才會拍板定案。你在 Task 4.1 中已經看過安排會面的過程了。在 Task 4.2 中，你將學到此過程中最常使用的字串。現在請做下列的 Task，記得做完後再核對答案！

Task 4.2

請將下列字串分門別類，並填入下表中。

- ... is good for me.
- ... is no good for me.
- ... is not OK for me.
- ... is OK for me.
- At this point I'd like to propose that + n. clause ...
- At this point I'd like to propose n.p. ...
- Can you make n.p. ...?
- Can you manage n.p. ...?
- How about Ving ...?
- How about n.p. ...?
- How is ... for you?
- I can't make it then.
- I can't manage it then.
- I'd like to suggest n.p. ...
- I'd like to suggest that + n. clause ...
- I'll be free at ...
- I'm afraid I have to V ...
- I'm afraid I'm going to be Ving ...
- I'm afraid I'm going to be ...
- I'm afraid I'm tied up then.
- I'm afraid that's not possible.
- Is ... OK for you?
- Let's V ...
- My schedule is really tight.
- My schedule is really tight + time.
- OK.
- Sure.
- That would be fine.

- That's fine.
- That's good for me.
- We could V ...
- What about Ving ...?
- What about n.p. ...?
- What time would suit you?

- When would be convenient for you?
- Why don't we V ...?
- Would ... be OK for you?
- Yes, I can make it.
- Yes, I can manage that.

提議會面 Making a Suggestion	拒絕提議 Rejecting a Suggestion	接受提議 Accepting a Suggestion

請繼續往下做 Task 4.3，暫時先不要核對上一題的答案。

Task 4.3

上列字串中那些語氣比較正式？那些比較不正式？請在語氣較正式的字串旁寫一個 F。

Task 4.2 & 4.3　參考答案

請利用下列語庫核對答案，並花一些時間研讀語庫。

Email 必備語庫　4.1

非正式 INFORMAL		
提議會面 Making a Suggestion	拒絕提議 Rejecting a Suggestion	接受提議 Accepting a Suggestion
• Can you make n.p. ...? • Can you manage n.p. ...? • How about Ving ...? • How about n.p. ...? • How is ... for you? • Is ... OK for you? • Let's V ... • We could V ... • What about Ving ...? • What about n.p. ...? • Why don't we V ...? • Would ... be OK for you?	• ... is no good for me. • ... is not OK for me. • I can't make it then. • I can't manage it then. • My schedule is really tight. • My schedule is really tight + time.	• ... is good for me. • ... is OK for me. • I'll be free at ... • OK. • Sure.

正式 FORMAL		
提議會面 **Making a Suggestion**	拒絕提議 **Rejecting a Suggestion**	接受提議 **Accepting a Suggestion**
• At this point I'd like to propose that + n. clause ... • At this point I'd like to propose n.p. ... • I'd like to suggest n.p. ... • I'd like to suggest that + n. clause ... • What time would suit you? • When would be convenient for you?	• I'm afraid I have to V ... • I'm afraid I'm going to be Ving ... • I'm afraid I'm going to be ... • I'm afraid I'm tied up then. • I'm afraid that's not possible.	• That would be fine. • That's fine. • That's good for me. • Yes, I can make it. • Yes, I can manage that.

💡 語庫小叮嚀

• 注意到了嗎？通常語氣比較正式的 set-phrases 比較長。

• 注意，propose 和 suggest 後面均是接名詞片語或「that + 名詞子句」。一般人常會寫成 suggest you to V 或 propose you to V，這是錯誤的。如果你經常犯這種錯誤，請盡量在讀本章時糾正此習慣。

• Can you manage Monday? 為英式用法；Can you make it on Monday? 為美式用法。

　　細節是非常容易出錯的地方，因此接下來的 Task 將訓練你更加注意 set-phrases 中的細節。記住，完成 Task 前，千萬不要先看答案。

📑 Task 4.4

請改正下列句子中 set-phrases 的錯誤，並將答案中的 set-phrases 畫底線。見範例 ①。

① At this point I'd like to propose us to meet next week.

→ *At this point I'd like to propose that we meet next week.*

② I'd like to suggest to cancel it.

→

③ I'd like to suggest we should postpone it.

→

④ Why don't we could try another day?

→

⑤ Can you manage to have our meeting on Monday?

→

⑥ How is next Wednesday to you?

→

⑦ We could to meet for lunch.

→

⑧ Would 3 pm is OK for you?

→

⑨ 4:30 pm it's no good for me.

→

⑩ 9 am was good for me.

→

⑪ I'm afraid I have to postponing the appointment.

→ _____

⑫ I'm afraid I'm tie up then.

→ _____

⑬ That's good to me.

→ _____

⑭ What time would suit for you?

→ _____

⑮ When would be convenient to you?

→ _____

Task 4.4 參考答案

② I'd like to suggest that we cancel it.

③ I'd like to suggest that we postpone it.

④ Why don't we try another day?

⑤ Can you manage Monday?

⑥ How is next Wednesday for you?

⑦ We could meet for lunch.

⑧ Would 3 pm be OK for you?

⑨ 4:30 pm is no good for me.

⑩ 9 am is good for me.

⑪ I'm afraid I have to postpone the appointment.

⑫ I'm afraid I'm tied up then.

⑬ That's good for me.

⑭ What time would suit you?

⑮ When would be convenient for you?

有些錯誤必須用心才找得出來。如果找不到錯誤，請仔細看答案或查看語庫。不要擔心這些 set-phrases 為何有誤，或是違反那些文法規則，只要專心研讀眼前的 set-phrases，背起來即可。挑幾個陌生、難度高或新的 set-phrases，花幾分鐘抄下來以加強記憶。抄好之後，請繼續做 Task 4.5。

Task 4.5

請利用語庫 4.1 的 set-phrases，填空並完成以下電子郵件。

①

> Dear Mary,
>
> We need to arrange a time to meet and discuss the new project. _____(A)_____ Thursday at 3 pm. I'm on leave for two days, but I'll be back on Thursday morning.

②

> Dear Joy,
>
> _____(B)_____ at the end of next week. _____(C)_____ Monday at 10 am?

③

> Dear Mary,
>
> Monday at 10 _____(D)_____, but I'll be free at 12. _____(E)_____ lunch?

④

> Dear Joy,
>
> Monday lunchtime _____(F)_____. I'll come to your office. See you then.

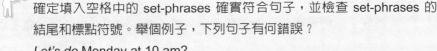

確定填入空格中的 set-phrases 確實符合句子，並檢查 set-phrases 的結尾和標點符號。舉個例子，下列句子有何錯誤？

Let's do Monday at 10 am?

沒錯，Let's V ... 的句尾不應該用問號。請再檢查一次你的答案，看看是否犯了類似的錯誤。記住，細節是最容易出錯的地方。

接下來請核對答案。

Task 4.5　參考答案

① Ⓐ I'd like to suggest
　　At this point I'd like to propose

---

② Ⓑ My schedule is really tight
　　I'm afraid I'm going to be out of town

　Ⓒ What about
　　How about

---

③ Ⓓ is no good for me
　　is not OK for me

　Ⓔ Can you make
　　Can you manage

---

④ Ⓕ is good for me
　　is OK for me

現在你已經做過一些 Task，加強了對安排會面 set-phrases 細節的注意，又學過在電子郵件中如何使用這些 set-phrases，現在就來練習寫電子郵件吧。請做接下來的延伸寫作題。這次的寫作題有兩個練習，一個讓你練習正式用語，一個讓你練習非正式用語。如果時間不夠，暫時先做一個練習。記住，做完練習後，才能看答案！

請閱讀下列摘要，然後看翻譯。看完翻譯後，請依摘要寫出五封往返的電子郵件。

> Mary and Tom are both middle managers in a big IT company.
> Because the company is so big, they don't meet often, and they do
> not know each other very well. Tom writes to Mary asking when they
> can meet to talk about the XYZ project. Mary suggests meeting at 4:00
> on Tuesday. Tom rejects this suggestion because he is going to be
> away all day on Tuesday. He suggests the following Thursday at 4 pm
> in meeting room 101.
> Thursday is good for Mary, but she rejects the afternoon because
> she has to visit the factory in the afternoon. She suggests meeting at
> 10:00 am instead. Tom accepts this suggestion.

翻譯

瑪莉和湯姆都是大型資訊科技公司的中階經理。由於公司很大，所以他們不常碰面，也不太認識對方。湯姆寫信給瑪莉問說，兩人什麼時候能見面談談 XYZ 的案子。瑪莉建議在星期二的四點見面。湯姆否決了這個建議，因為他星期二整天都不在。他建議在隔週的星期四下午四點到 101 號會議室。

瑪莉星期四可以，但她否決了，因為她下午必須去一趟工廠。她建議改到早上十點見。湯姆接受了這個建議。

①

> Dear Mary,
>
> I'd like to arrange a time to meet and talk about the XYZ project. When
> would be convenient for you?
>
> Yours,
> Tom

②

Dear Tom,

Thanks for your email. At this point I'd like to suggest 4 pm on Tuesday. I hope this is OK for you.

Yours,
Mary

③

Dear Mary,

I'm afraid that's not possible. I'm going to be away all day on Tuesday. Would Thursday at 4 pm in meeting room 101 be OK for you?

Tom

④

Dear Tom,

Thursday is good for me, but I'm afraid I'm going to be visiting a factory in the afternoon. Would the morning at 10:00 be OK for you?

Mary

⑤

Dear Mary,

That's good for me. I'll see you then.

Tom

請閱讀下列摘要，然後看翻譯。看完翻譯後，請依摘要寫出五封往返的電子郵件。

> Denni and Carrie work in different companies in the same industry. They used to work in the same company and know each other very well. They are good friends. Denni writes to Carrie asking when they can meet to discuss job opportunities in Carrie's company, as Denni wants to change her job. Carrie suggests a meeting at 5 pm on Thursday. Denni rejects this suggestion because she has to take her daughter to the dentist at that time. She suggests Friday at 6 at Starbucks instead. Friday is good for Carrie, but she rejects 6 because she will be in a meeting with her boss until 7. She suggests they meet at 8 and have dinner together. Denni accepts this suggestion.

翻譯

丹妮和凱莉從事同一行業，但在不同公司上班。他們以前是同事，彼此是很熟的好朋友。丹妮想換工作，因此寫信詢問凱莉何時可以見面，聊聊凱莉公司有什麼職缺。凱莉建議星期四下午五點見面，但是丹妮那時必須帶女兒去看牙醫，所以拒絕了，她提議星期五晚上六點在星巴克見。星期五凱莉有空，但是當天要和老闆開會到晚上七點，因此六點趕不及。她建議八點時一起吃晚餐。丹妮接受建議。

延伸寫作題 4.1B 參考答案

①

> Dear Carrie,
>
> I'd like to talk to you about job opportunities at your company. I'm really fed up here. When is a good time for us to get together?
>
> Denni

②

Dear Denni,

Sorry to hear you are fed up with your job. Let's meet. Is 5 on Thursday OK for you?

Carrie

③

Dear Carrie,

I can't make it then. I'm taking my daughter to the dentist. What about Friday evening after work? 6 at Starbucks? I'll buy you a coffee.

Denni

④

Dear Denni,

I can't manage it so early. I'm going to be in a meeting with my boss until 7. Can you make 8? We could have dinner and start the weekend together.

Carrie

⑤

Dear Carrie,

Sure. I'll see you then.

Denni

請回到 Task 4.1，比較這次和先前畫了底線的字串。你進步了多少？

接下來的這一節中，我們要學的是描述和更改會面時間、地點的字串。讀本節時，請記住所有的「安排」都是指未來的事。現在請做下面的 Task。

📖 **Task 4.6**

請閱讀下列電子郵件，找出描述會面的字串，並在底下畫直線；找出更改會面時間或地點的字串，並在底下畫曲線。

①

Dear Joy,

Just a quick note to confirm our meeting on Monday at 12. Let's meet downstairs in the lobby of your building. I'll be bringing my assistant to help us with some of the data. I'm meeting a client at 2, so I hope the meeting can be over by 1:30. Not much time, I know, but my schedule is pretty tight next week. Anyway, I'll see you then. Have a nice weekend.

Mary

💡 **Email 摘要**

瑪莉向喬伊確認星期一中午十二點開會，並建議在喬伊辦公大樓的大廳見。她告訴喬伊，她將帶助理來幫忙處理資料，不過下午兩點和一名客戶有另外的會議要開，所以希望下午一點半前結束會議。

Vocabulary

lobby [ˈlɑbɪ] *n.* 大廳

②

Dear Julie,

We are going to have to cancel our meeting next week, I'm afraid. Michael wants me to conduct a spending review of the department. However, I'm having lunch with Mary on Wednesday so I can ask her for her input and report back to you on the situation. I'm also seeing John at the Director's Evening tonight so I'll ask him for his ideas too. Sorry once again. I'll be in touch with you tomorrow and we can reschedule our meeting.

Clinton

♀ Email 摘要

柯林頓寫信告訴茱莉，老闆麥可要他做部門支出報告，下星期和茱莉的會議必須取消。他說星期三和瑪莉共進午餐時會詢問她的想法，然後再轉告茱莉；也會在今晚的主管晚會上詢問約翰的意見。他說明天會和茱莉聯絡，重新安排會議時間。

　　我們稍後再回來討論這兩封電子郵件。Task 4.7 中有描述和更改未來會面時間、地點的常見字串。現在請做 Task，做完後再檢查答案！

📄 Task 4.7

請將以下字串分門別類，並填寫於下列表格中。

- ... be going to V ...
- ... be Ving ...
- I need to change/cancel/postpone n.p. ...
- I'll be arriving ...
- I'll be leaving ...
- I'll have to V ...

- I'm afraid I have to V ...
- I'm afraid we're going to have to cancel n.p. ...
- I'm afraid we're going to have to postpone n.p. ...
- I'm going to have to V ...
- I've got a meeting ...
- I'll be Ving ...
- I've got an appointment ...
- Would it be OK if we postponed it till ...?

描述會議相關事項 Describing	更改會議相關事項 Changing

Task 4.7 參考答案

Email 必備語庫 4.2

描述會議相關事項 Describing	更改會議相關事項 Changing
... be going to V ...	I need to change/cancel/postpone n.p. ...
... be Ving ...	I'm afraid I have to V ...
I'll be arriving ...	I'm afraid we're going to have to cancel n.p. ...
I'll be leaving ...	
I'll be Ving ...	I'm afraid we're going to have to postpone n.p ...
I'll have to V ...	
I've got a meeting ...	I'm going to have to V ...
I've got an appointment ...	Would it be OK if we postponed it till ...?

語庫小叮嚀

- 你可能已經認識語庫中的一些字串，例如 be going to V 和現在進行式。這些字串在電子郵件中是用來描述未來的安排。
- 注意 I'll be leaving ...、I'll be arriving ... 和 I'll be Ving ...，而非 I will leave ... 等。
- 另外，注意語庫中沒有 I will V，這是因爲 I will V 不表示未來的安排。在本章「語感甦活區」，我會討論 will 的用法。但現在請先回想一下，描述安排事項時，你是否經常用錯 I will V 呢？

挑一些陌生、困難或新的 set-phrases，花幾分鐘仔細抄寫下來以幫助記憶。完成後，請繼續做 Task 4.8。

Task 4.8

配合題，請為左欄中的字串，由右欄中找出適當的字串，以合併成完整的句子。然後屬於描述會面時間、地點的句子，標上 D；屬於更改會面時間、地點的句子，標上 C。見範例 ①。注意，答案可能不止一組。

1	I'm going	a	a meeting with the director.
2	I'm afraid I	b	have to drop out of the conference.
3	I've got	c	have to finish an urgent report by tomorrow.
4	I'm afraid we're going to have to	d	is arriving on CI642.
5	I'll be	e	postpone our meeting till next week.
6	I'm going to	f	seeing her tonight.
7	He	g	to be in Hong Kong next week.
8	Would it be OK if	h	we had our meeting later in the week?

① _____ *g* D _____   ④ _____   ⑦ _____
② _____   ⑤ _____   ⑧ _____
③ _____   ⑥ _____

現在請核對答案。記住,答案重點不在於文法,而是眼前的字串。

📋 **Task 4.8** 參考答案

② b C   c C   ⑥ a D   b C   c C   e C

③ a D   ⑦ d D

④ e C   ⑧ h C

⑤ f D

📋 **Task 4.9**

請利用語庫 4.2 的 set-phrases,填空並完成下列電子郵件。

①

> Dear Joy,
>
> Just a quick note to confirm our meeting on Monday at 12. Let's meet downstairs in the lobby of your building. _____Ⓐ_____ my assistant to help us with some of the data. _____Ⓑ_____ a client at 2, so I hope the meeting can be over by 1:30. Not much time, I know, but my schedule is pretty tight next week. Anyway, I'll see you then. Have a nice weekend.

②

> Dear Julie,
>
> _____Ⓒ_____ our meeting next week, I'm afraid. Michael wants me to conduct a spending review of the department. However, _____Ⓓ_____ lunch with Mary on Wednesday so I can ask her for her input and report back to you on the situation. _____Ⓔ_____ John at the Director's Evening tonight so I'll ask him for his ideas too. Sorry once again. I'll be in touch with you tomorrow and we can reschedule our meeting.

① Ⓐ I'm going to be bringing

　　I'm bringing

　　I'll have to bring

　Ⓑ I've got a meeting with

　　I've got an appointment with

　　I'm going to meet

　　I'll be meeting

　　I'm going to be meeting

　　I'll have to meet

② Ⓒ I need to cancel

　　I'm afraid I have to cancel

　　I'm afraid we are going to have to postpone/cancel

　　I'm going to have to postpone/cancel

　Ⓓ I'm going to have

　　I'm having

　　I'll be having

　　I'll have to have

　Ⓔ I'll also be seeing

　　I'm also going to see

　　I'm seeing

　　好，現在該是再做一個延伸寫作題的時候了。目的是讓你練習使用確認會面的字串。記住，練習完成後才能檢查答案喔！

請閱讀下列摘要，然後看翻譯。看完翻譯後，請依摘要寫一封電子郵件。

Michael and Tien are quite good friends. They both work for a furniture manufacturer. Tien is writing to Michael to postpone their appointment next week, because he has to visit the factory to help them understand the specs for a new product. He tells Michael that arranging another appointment in this month is going to be difficult because he is going to Texas. He also tells Michael that he has been relocated to Texas, starting from the end of next month, so he is going to be busy looking for somewhere to live there. He suggests that they try to arrange an appointment on the phone on Monday.

翻譯

麥可和田是相當好的朋友。他們都在一家家具製造廠上班。田正在寫信給麥可，將下星期的會面延期，因為他必須去工廠協助他們搞懂新產品的規格。他告訴麥可，這個月要安排另一次會面有困難，因為他要去德州了。他還告訴麥可，從下個月底開始，他就會搬到德州去，所以他要忙著在那裡找地方住。他建議彼此在星期一通個電話，以設法安排會面。

現在請以下面的參考答案核對你寫的郵件。

Vocabulary

manufacturer [ˌmænjəˈfæktʃərə] *n.* 製造業者；廠商
relocate [reˈloket] *v.* 遷移；重新安置

Dear Michael,

I need to cancel our appointment next week. They are having problems understanding the specs for the new product at the factory, so I'm afraid I have to visit them to help them with that. I don't know when we can reschedule the appointment, as next month I'm going to be really busy. I'm flying to Texas.

Also, I should tell you that I am relocating to Texas next month, so I'll be looking at apartments there.

Anyway, let's talk on Monday and see if we can arrange another time to meet.

Best regards,
Tien

請回到 Task 4.6，看看你在當中畫了底線的字串。你進步了多少呢？

對於以中文為母語的人來說，最常用錯的英文字可能就是 will 了，經常以為 will 即表示未來。但是未來的事情可分成兩種，用英文表示時，必須慎重區分：個人的事情（例如個人或認識之人的計畫和安排）和非個人的事情（例如對於經濟、國家或政治等情況之未來的預測或說法）。

Task 4.10

請將下列句子分門別類，並填入下表中。

- Exports will be affected.
- I'll see you tomorrow.
- I'm going to Thailand next week.
- Market share will stay roughly the same.
- Mary is having a baby!
- SARS is going to have a negative effect on growth.
- The economy is expected to shrink by 2%.
- The NT dollar will probably weaken.
- We'll be staying at the Hyatt.
- When are you going to New York?

Personal Future	Impersonal Future

請以下表核對答案。

Personal Future	Impersonal Future
• I'll see you tomorrow.	• Exports will be affected.
• I'm going to Thailand next week.	• Market share will stay roughly the same.
• Mary is having a baby!	• SARS is going to have a negative effect on growth.
• We'll be staying at the Hyatt.	• The economy is expected to shrink by 2%.
• When are you going to New York?	• The NT dollar will probably weaken.

　　你是否清楚為何有些句子屬於個人事情，有些屬於非個人事情？如果不確定，請看各句子的主詞。在個人事情欄中，所有的主詞都和個人或認識的人有關；非個人事情欄中，所有的主詞都與個人事物或情況無關。未來的事情有個人及非個人之分，能夠區分這兩者是非常重要的。

　　當 will 用於個人事情中，含有即時性的隱意。一般而言，任何會面都是事先經過安排、計畫和周詳考慮的結果，不會是即時性的，因此對會面的描述不應使用 will。但是電子郵件中，某些含 will 或 'll 的 set-phrases 經常出現，最好的辦法是將其視為功能性的 set-phrases，而非意義性的。這些 set-phrases 的功能包括：提議、做承諾、提出請求和即時做決定等，其共通點是均為即時性，而且與會面無關。請做接下來的 Task，同時思考一下這種 set-phrases 的特性。

Task 4.11

請將下列字串分門別類，並填入下表中。

- I'll get it.
- I'll give you a ring.
- I'll do it.
- I'll V ...
- I'll drop you a line.
- I'll send it tomorrow.
- I'll be free at ...
- I'll be in touch.
- I think I'll probably V ...
- Shall I do it?
- Will you do it?
- I'll get back to you as soon as I can.

- I'll be back on Thursday.
- She'll be back tomorrow.
- I'll be late.
- I'll see you then.
- It'll be alright.
- It'll take time.
- We'll see.
- Will you help me with this?
- I'll see what I can do.
- I think X will V ...
- I think it'll V ...

回應請求 Making Offers	請求協助 Making Requests
提出承諾 Making Promises	其他 Other

請利用下列語庫核對答案。

Email 必備語庫 4.3

回應請求 Making Offers		請求協助 Making Requests
• I'll get it.	• I'll do it.	• Will you do it?
• I'll V ...	• Shall I do it?	• Will you help me with this?
• I'll see what I can do.		
提出承諾 Making Promises		**其他 Other**
• I'll give you a ring.	• I'll V ...	• It'll be alright.
• I'll drop you a line.	• I'll be in touch.	• It'll take time.
• I'll send it tomorrow.		• We'll see.
• I'll be free at ...		• I think I'll probably V ...
• I'll get back to you as soon as I can.		• I think X will V ...
• I'll be back on Thursday.		• I think it'll V ...
• She'll be back tomorrow.		• I'll be late.
• I'll see what I can do.		
• I'll see you then.		

💡 **語庫小叮嚀**

- 如果你的答案與參考答案相距甚大，別訝異或失望。含 will 的 set-phrases 有一個問題，就是除非知道是在什麼情境下使用，否則往往很難看出它的 實際功能。你只要知道這些 set-phrases 的功能是即時性的就可以了。
- 有些 set-phrases 為慣用語：
  - *I'll give you a ring.* 「我會打電話給你。」
  - *I'll drop you a line.* 「我會寫信給你。」
- Shall 通常用在問句中，表示提議。

　　請回去看本章中所有的電子郵件，找出含有 will 的 set-phrases 並畫底 線。這些 set-phrases 的功能為何？請花幾分鐘研究一下，並將它們按照語庫 4.3 分門別類。

Unit
4

請閱讀以下電子郵件，由題目下方的選項挑出符合文意的字串，並填入空格中。見範例①。

①

> Dear Lisa,
>
> The client can't come in the morning, so Ⓐ <u>I'll move the meeting to</u> the afternoon. Can you lend me your notes on the project, as I have lost mine. Ⓑ <u>I'll give them back to you</u> after the meeting.
>
> Ⓐ I've got a meeting in/I'll move the meeting to
> Ⓑ I'm going to give them back to you/I'll give them back to you

②

> Dear Mark,
>
> Thanks for your email asking me for some data for your report. I think this data might be a little difficult to collect, but Ⓒ_____.
>
> Ⓒ I'll be seeing what I can do/I'll see what I can do

③

> Dear Cora,
>
> Regarding your question about the timeframe for the project, I don't know myself. Ⓓ_____ know. I've got a meeting with him tomorrow afternoon, so Ⓔ_____ then, and Ⓕ_____ later.
>
> Ⓓ I think John is going to/I think John will
> Ⓔ I'm going to ask him/I'll ask him
> Ⓕ I'm going to drop you a line/I'll drop you a line

④

> Dear Jonathon,
>
> I have been asked to provide a report on your department for the regional head office. ⓖ_____ with it? I need your expertise. They have given me a list of questions they want to see answered in the report. ⓗ_____ to you tomorrow.
>
> ⓖ Are you helping me/Will you help me
> ⓗ I'm going to send it/I'll send it

📖 **Task 4.12** 參考答案

② ⓒ I'll see what I can do

③ ⓓ I think John will
   ⓔ I'll ask him
   ⓕ I'll drop you a line

④ ⓖ Will you help me
   ⓗ I'll send it

　　希望你全都答對了。最後一題中兩個字串皆可用，前者表示寄件者決定以其他方式寄出文件（例如用 DHL 國際快遞），後者可表示寄件者所做的決定。記住：**文法就是意義，意義就是選擇！**

　　請拿出最近寫的電子郵件，找出用錯 will 的地方，並檢查對會面和未來事情的寫法是否有錯。改正其中的錯誤，下次寫電子郵件時，避免重蹈覆轍。

　　現在請回到本章前言中設定的學習目標，勾出已經學會的項目。如果還有尚未學會的項目，過幾天後重新研讀本章，或許便會看懂了。另外，記住要大量閱讀英文母語人士所寫的電子郵件，別小看閱讀對加強英文的效果！

Unit 5

解決問題的 Email

　　本章中，我們要學一種常見的電子郵件類型——解決問題。很多人都有過這樣的經驗：接獲客戶、消費者或同事的投訴或抱怨，或者請求解決問題，因此清楚地寫出解決問題的辦法，是不可或缺的寫作技巧。換一個角度來看，你也可能會遇到問題，需要他人幫忙解決，因此清楚描述問題，爭取適當協助，也是寫作時必備的本事。

　　本章要教的便是這兩種寫作技巧。但是不同於本書其他章節著重 set-phrases，本章的教學將以 chunks 爲主。還記得我在＜前言＞中教過的 chunks 嗎？ Chunks 是固定和流動元素的組合，通常意義字和功能字兼具。如果不太記得，可以溫習一下＜前言＞中 chunks 的定義。至於本章的電子郵件範例，均將以電腦軟體程式的問題爲主，原因有四：

• 我不知道你從事哪一個行業，無從猜測你在工作上可能遇到的問題。
• 本書讀者來自各行各業，每個行業在工作上遭遇的問題也不同。
• 上班族對基本的文書處理軟體（word processor）和試算表程式（spreadsheet program）都不陌生。
• 雖然是以此類軟體程式的問題和解決方法爲範例，但其中用到的 chunks 其他行業仍然適用，我也會教你這些 chunks 的用法。

　　如果你沒有寫這類電子郵件的需要，跳過本章直接看下一章也無妨。但我在本章中提到的一些字串極爲常見，一般人又時常用錯，建議你還是研讀本章，或許能夠進一步加強寫作。本章一開始，請先做第一個 Task。

Task 5.1

請閱讀下列兩封電子郵件，找出描述或解決問題的字串，並畫底線。

Dear IT Department,

I'm having problems getting on to the Net. Whenever I click the internet icon on my toolbar, nothing happens. None else in the office is having this problem, and I can still send and receive email. What should I do?

Mary

Ⓐ

Dear Mary,

Sorry to hear you are having problems. I can think of two reasons why you may be having this problem. First, your Internet settings might not be correct. To check your settings, double-click the Internet icon in the Control Panel, and then make any corrections needed.

Secondly, you may be having problems with your modem. Make sure your modem is working correctly. For more information about how to do this, see the topic on modem troubleshooting in the help file.

If you have any more problems with this, please call me and I'll come over and see if I can find out what's wrong.

IT Department

Unit
5

Vocabulary

click [klɪk] *v.* 移動游標並按滑鼠上的按鈕

toolbar [ˋtulˌbɑr] *n.* 電腦螢幕上的工具列

modem [ˋmodəm] *n.* 數據機

icon [ˋaɪkɑn] *n.* 圖像

Control Panel 控制面板

troubleshooting [ˋtrʌblˌʃutɪŋ] *n.* 解決問題

　　我們等一下會回來討論這兩封電子郵件，你可以在本章最後一節檢查答案。現在請先看一看本章的學習目標吧。看完本章，你應該達到的學習目標如下：

☐ 準確描述問題，適當地請求協助。

☐ 清楚寫出問題來源，提出建議和說明解決步驟。

☐ 描述問題的可能來源，並針對各來源提出解決方法與各方法的結果。

☐ 練習運用本章教的字串，寫出關於軟體程式的問題。

☐ 學會運用本章教的字串，描寫不同行業中發生的問題。

☐ 練習運用本章教的字串，描寫自己行業和工作上遇到的問題。

☐ 認識新字串——problem 和 solution。

　　好，我們這就進入本章的第一節吧！

描述問題與請求協助

本節中，首先要教的是精準描述問題和請求協助的字串。開始學習之前，來做一個學前測驗吧！

Task 5.2

請閱讀下列摘要，根據其中一個摘要寫電子郵件，描述問題並請求協助。

① 電腦的數據機壞了。請求協助。
② 昨天寫的檔案不見了。請求協助。
③ 詢問取消列印工作的方法。
④ 你得將報告存入隨身碟寄給客戶，但是隨身碟無法存檔。
⑤ 文書處理軟體上的繪圖工具列不見了，詢問叫出工具列的方法。

Task 5.2　參考答案

① My modem won't work. Please help.
② I can't find a document I created yesterday. Any suggestions?
③ How can I cancel the print job once it's started printing?
④ I'm having problems saving my work onto a USB drive. Any advice you can give would be really appreciated as I have to send the USB drive to my client soon.
⑤ The drawing toolbar on my spreadsheet program seems to have disappeared. What can I do to get it back?

Unit
5

注意到了嗎？所有參考答案的模式如出一轍：先描述問題，再請求協助。

Vocabulary

document [ˈdɑkjəmənt] *n.* 公文；文件　　　　**USB drive** 隨身碟

我們來看看這些參考答案中用到的字串吧。現在請做下面的 Task。你可以回頭借助 Task 5.2 的電子郵件，但是先別看參考答案喔！

📑 Task 5.3

請將下列 chunks 和 set-phrases 分門別類，並填入下表中。

- How do you ...?
- How does it ...?
- How does X work?
- I can't V ...
- I'm having problems Ving ...
- I'm having problems with n.p. ...
- I'm having trouble Ving ...
- I'm having trouble with n.p. ...
- Any advice you can give would be greatly appreciated.
- Any advice?
- Any suggestions?
- ... be giving me problems.
- ... be giving me trouble.
- ... doesn't work.
- ... doesn't work properly.
- ... doesn't work very well.
- ... doesn't work when(ever) + n. clause ...
- How can I ...?
- How do I ...?
- I seem to have p.p./n.p. ...
- I can't seem to V ...
- Is it possible to V ...?
- ... doesn't seem to V ...
- ... doesn't seem to be Ving ...
- ... seems to V ...
- It seems that + n. clause ...
- ... seems to have p.p./n.p. ...
- Please help.
- What can I do?
- What should I do?
- ... won't work.
- ... won't work properly.
- ... won't work when(ever) + n. clause ...
- Whenever I V ..., nothing happens.
- Whenever I V ..., it V ...
- Please advise.

描述問題 Describing Problems	請求協助 Asking for Help

Task 5.3 參考答案

請利用下列語庫核對答案。

描述問題 Describing Problems	請求協助 Asking for Help
• I can't V ...	• What can I do?
• I seem to have p.p./n.p. ...	• What should I do?
• I'm having problems Ving ...	• How do you ...?
• I'm having problems with n.p. ...	• How does it ...?
• I'm having trouble Ving ...	• How does X work?
• I'm having trouble with n.p. ...	• How can I ...?
• ... be giving me problems.	• How do I ...?
• ... be giving me trouble.	• Any advice you can
• ... doesn't work.	give would be greatly
• ... doesn't work properly.	appreciated.
• ... doesn't work very well.	• Any advice?
• ... doesn't work when(ever) + n. clause ...	• Any suggestions?
• ... won't work.	• Is it possible to V ...?
• ... won't work properly.	• Please help.
• ... won't work when(ever) + n. clause ...	• Please advise.
• I can't seem to V ...	
• ... doesn't seem to V ...	
• ... doesn't seem to be Ving ...	
• ... seems to V ...	
• It seems that + n. clause ...	
• ... seems to have p.p./n.p. ...	
• Whenever I V ..., nothing happens.	
• Whenever I V ..., it V ...	

💡 **語庫小叮嚀**

• 注意句子中哪些 set-phrases 放在句首，哪些放在句尾。

• 注意主詞：有時是 I，有時是 It。主詞不可易位。

• 注意所有可能用到的結尾：V、n. clause、n.p.、Ving 或 p.p.。

現在請回頭看 Task 5.2 的參考答案，找出參考答案中屬於必備語庫的字串，並注意這些字串的用法。你可能覺得有兩處特別難，一個是 set-phrases 的結尾，一個是用對主詞。現在請做下一個 Task，千萬不要先看答案！

Task 5.4

請填空，完成下列電子郵件。

① My modem _____. Please help.

② _____ find the document I made yesterday. Any suggestions?

③ _____ stop the printer once it's started printing?

④ _____ saving my work onto a USB drive. Any advice you can give would be really appreciated, as I have to send the USB drive to my client.

⑤ The drawing toolbar on my spreadsheet program _____ disappeared. What can I do to get it back?

Task 5.4　參考答案

Unit
5

① won't work
　won't work properly
　doesn't work
　doesn't work properly
　doesn't work very well
　is giving me problems
　is giving me trouble

② I can't
　I can't seem to

③ How can I
　How do you
　How do I
　Is it possible to

④ I'm having trouble
　I'm having problems

⑤ seems to have

現在可以回去看看你在 Task 5.2 中寫的電子郵件，是否有須修正之處？修改完後，請繼續做下一個 Task，運用方才學的這些字串寫電子郵件。

Task 5.5

請從 Task 5.2 選出另一個摘要，根據該摘要寫一封電子郵件。

Task 5.5 參考答案

見 Task 5.2 之參考答案。

現在你已經學會這些字串的用法，能夠向他人請求幫忙、解決軟體程式的問題了。接下來可以學習如何用這些字串描述自己行業中遇到的問題。

下一個 Task 中，有兩封來自不同行業的電子郵件範例。做 Task 時，請找出學過的字串、畫底線，並思考在不同行業中這些字串的用法。

Task 5.6

下面有兩封電子郵件，分屬於兩個不同行業，請寫下這些電子郵件所屬的行業。

①

Dear Mary,

I'm afraid I am having trouble with the last shipment. It seems that the documentation which was sent with the shipment refers to a different shipment. For this reason the goods have been impounded by customs, and they will not release them until they have the correct documentation. Whenever I try to explain to the customs officers what has happened, they refuse to listen.

What shall I do? Any advice you can give would be greatly appreciated.

Mike

②

Dear George,

The new accounting procedure you outlined last month doesn't work well for us. It may be OK for a smaller company, but for an accounting department in a financial company as big as ours, it simply won't work. There are too many transactions and the transactions are all quite complex. Your system is giving us more problems than we had before.

Is it possible for you to either explain the new system again, or to allow us to go back to the old way of doing things?

Please advise.

Gina

💡 Email 摘要

1. 寄件者表示，由於貨品的文件有誤，導致海關扣住了貨品，他請瑪莉幫忙解決問題。
2. 寄件者表示，她不會用新的會計系統，並覺得公司的交易金額和類型也不適合用這個新系統，請喬治幫忙解決此問題。

📑 Task 5.6 參考答案

① 進出口業（Import/Export）　　② 金融業（Finance）

　　再回頭看一遍這些郵件吧。找出目前學過的字串，並留意在不同行業中這些字串的用法，這對未來自己在工作時會很有幫助。接著，試著以自身工作時常面臨的問題來寫封郵件，你應該能隨心運用這些字串了才是！

Vocabulary

transaction [trænˋzækʃən] *n.* 業務；交易

　　本節中，我們要學的是描述解決方法的字串，這些字串可分成幾類。開始學習之前，我們來看一封解決問題的電子郵件吧。現在請做 Task 5.7。

Task 5.7

請看回覆問題的電子郵件，找出敘述解決問題方法的字串，並畫底線。

Q

> I have lost a document I made yesterday. What should I do?

A

> You may have saved it in the wrong place. Have you tried using the search function to look for it? To use the search facility, click on the Start icon, then click Search. Select the second item on the menu: Documents, then enter the name of the document and click Search.
>
> If you can't remember the name of the document, you can click the Start icon, and then click My Recent Documents and see if you can find it on the list. Check first that you can remember the right name of the document! Alternatively, you might have accidentally deleted it. Why don't you look in the recycle bin? Make sure you have not moved it there by mistake. To check the recycle bin, click the Recycle Bin icon on your desktop.
>
> I hope this helps.

Vocabulary

alternatively [ɔlˋtɜnəˌtɪvlɪ] *adv.* 或者；二者擇一地

172

Q. 寄件者的檔案不見了，請求協助。

A. 對方回信說問題有兩種可能性，一個是檔案可能存錯地方，另一個是可能丟到資源回收桶中了。他解釋如何用搜尋功能或從最近使用的文件中找檔案，以及開啟資源回收桶的方法。

我們等一下會回來討論這封電子郵件，請到時再檢查答案。不過請注意一點，**電子郵件中第一件事便是幫對方**找出問題來源：

You may have saved it in the wrong place.
You might have accidentally deleted it.

第二件事是提出建議：

Have you tried using the search function to look for it?
Why don't you look in the recycle bin?

或者說明解決步驟：

Make sure you have not moved it there by mistake.
Check first that you can remember the right name of the document!

請回頭看本章 Task 5.1 的電子郵件，你是否看出同樣的模式？如果看不出來，別擔心，保證你待會兒就能看出。現在我們來進一步看字串的細節，請做下面的 Task。

Task 5.8

以下為解決問題時會使用的 chunks 和 set-phrases，請分門別類，並填入下表中。

Unit
5

- ... check n.p. ...
- ... check that + n. clause ...
- ... try Ving ...
- ... try again later.
- You might have forgotten to V ...
- You could try Ving ...
- What about Ving ...?
- How about Ving ...?
- Make sure that + n. clause ...
- ... may (not) be ...
- ... might (not) be ...

- Let's V ...
- ... could be ...
- ... may (not) have p.p. ...
- ... might (not) have p.p. ...
- ... could have p.p. ...
- You may have forgotten to V ...
- Why don't you/we V ...?
- Have you tried Ving ...?
- Have you remembered to V ...?
- See if you can V ...

找出問題 Finding the Cause of the Problem	
建議解決方式 Suggesting a Solution	進一步建議 Recommending Action

 Task 5.8 參考答案

請利用下列語庫核對答案。

找出問題 Finding the Cause of the Problem	
• ... check n.p. ...	• ... may (not) have p.p. ...
• ... check that + n. clause ...	• ... might (not) have p.p. ...
• You might have forgotten to V ...	• ... could have p.p. ...
• ... may (not) be ...	• You may have forgotten to V ...
• ... might (not) be ...	• Have you tried Ving ...?
• ... could be ...	• Have you remembered to V ...?

建議解決方式 Suggesting a Solution	進一步建議 Recommending Action
• You could try Ving ...	• ... check n.p. ...
• What about Ving ...?	• ... check that + n. clause ...
• How about Ving ...?	• ... try Ving ...
• Why don't you/we V ...?	• ... try again later.
	• Make sure that + n. clause ...
	• Let's V ...
	• See if you can V ...

💡 **語庫小叮嚀**

• 注意 may be、might be、could be 後面均會接 Ving 或形容詞。

• 注意，雖然 may be 和 might be 可以寫成否定式 may not be 和 might not be，但是這裡的 could be 不行。

• 注意 may have、might have、could have 後面是接 p.p.。

• 注意 check 後面可以接「名詞片語」，或「that ＋名詞子句」。

Unit
5

　　語庫的 Finding the Cause of the Problem（找出問題的原因）一欄中，請注意可以選擇用 may be 或 may have p.p.、might be 或 might have p.p.、could be 或 could have p.p.。這些 chunks 有什麼差別呢？這問題需要更詳細地解釋，答案可能和你對這類動詞—— modal verbs（語態動詞）的認識有著天壤之別，請做好心理準備。基本上，這些 chunks 的差別在於：

may (not) be might (not) be could be	均指現在 引起問題 的原因	• Your modem <u>may not be</u> working. • Your settings <u>could be</u> wrong.
may (not) have p.p. might (not) have p.p. could have p.p.	均指過去 引起問題 的原因	• You <u>may have forgotten</u> to turn on your printer. • You <u>might have saved</u> the document under a different name. • You <u>could have</u> accidentally <u>put</u> the document into the recycle bin.

　　如果你覺得這樣的說明難以理解，請重新再看一遍。若仍然感到困惑，別硬逼自己看懂，只要接受「**may be** ＝現在原因」；「**may have p.p.** ＝過去原因」就好。接下來的 Task 或許有助於理解。

📖 Task 5.9

表示現在原因的句子，請在空格填上 n，表示過去原因的句子，請填上 p。見範例①。

____p____ ① You might have forgotten to turn on the power.

_____ ② The cartridge could have run out of ink.

_____ ③ The software package might not be compatible with your operating system.

_____ ④ The server could be down.

_____ ⑤ The settings may have been changed.

_____ ⑥ There may not be enough memory to run the program.

_____ ⑦ You may have forgotten to pay your subscription.

_____ ⑧ The paper might have jammed.

📖 Task 5.9　參考答案

② p　　③ n　　④ n　　⑤ p　　⑥ n　　⑦ p　　⑧ p

176

為加強對這部分的學習，請回到本章第一封電子郵件—— Task 5.1，研讀其中屬於這類的字串。看完後，再繼續看本章。

Task 5.10

請回去看 Task 5.2 的參考答案，挑選其中一個遇到的問題，嘗試用學過的字串寫一封回信說明解決的方法。

請研讀下列參考答案，進一步增加學習效果。

Task 5.10　參考答案

① Q: My modem won't work. Please help.

　A: Your modem may not be compatible with your operating system. Have you tried checking your settings?

② Q: My modem won't work. Please help.

　A: You may have forgotten to plug in the phone line. Check that your connection is good.

③ Q: My modem won't work. Please help.

　A: Your modem card could be in the wrong slot. Make sure that your modem card is in the right port.

好，現在我們來學習用於解決問題的另一套字串——描述不同的解決方法和後果。在 Task 5.1 和 5.7 中，你已經注意到寄件者如何描述問題的可能原因：

If you can't remember the name of the document, ...
To check your settings, ...

Vocabulary

slot [slɑt] *n.* 溝槽；位置　　　　　**port** [port] *n.* 連接埠

然後提出幾個解決方法或後果：

... you can click the Start icon, ...

... double-click the Internet icon in the Control Panel, ...

現在來深入看一看這類字串。請看下表，左欄是用來表示問題的可能情況，右欄是用來表示左欄中可能問題的動作（actions）、選擇（options）或結果（consequences）。表格下方為左右欄合併後組成的幾個例句。

If you don't want to V, ...	you should V ...	it ought to V ...
If you want to V, ...	it should V ...	you will V ...
To V, ...	you could V ...	it will V ...
For n.p., ...	it could V ...	you might V ...
If it V, ...	you can V ...	it might V ...
If it doesn't V, ...	you must V ...	V ...
	you ought to V ...	you need to V ...

- 〔動作〕To save your work in a safer format, you should select the Rich Text Format.
- 〔選擇〕For easier access to your files, you could store them on your desktop.
- 〔結果〕If you right click the mouse, you will see a menu on the screen.

現在輪到你試一試身手了。

Task 5.11

請將上表右欄中的字串分門別類，並填入下表中。

- ... you should V ...
- ... it should V ...

- ... you could V ...
- ... it could V ...

- ... you can V ...
- ... you must V ...
- ... you ought to V ...
- ... it ought to V ...
- ... you will V ...

- ... it will V ...
- ... you might V ...
- ... it might V ...
- ... V ...
- ... you need to V ...

建議方式 Options	解決方式 Actions	描述結果 Consequences

Task 5.11 參考答案

Unit
5

請利用下列語庫核對答案。

Email 必備語庫 5.3

建議方式 Options	解決方式 Actions	描述結果 Consequences
... you could V ... ... you can V ... ... you might V ...	... you should V ... ... you must V ... ... you ought to V ... ... V ... ... you need to V ...	... it should V ... ... it could V ... ... it ought to V ... ... you will V ... ... it will V ... ... you might V ... ... it might V ...

請閱讀下列電子郵件，找出描述解決方法和結果的字串，並畫底線。為 if 句子中的第一個子句底下畫直線，為第二個子句底下畫曲線。

If you want to merge cells, highlight the cells you want to merge, click Table on the menu bar, and then click Merge Cells. If you want to reverse this process, you should highlight the cells you want to split and then choose Split Cells from the Table menu. If this doesn't work, you need to make sure you have highlighted the cells first. If you want to adjust the size of your cells, you could place the cursor over a line and click until a dotted line appears and then drag. Alternatively, you can also adjust the cell size by highlighting the table, selecting Properties from the Table menu, and then selecting Cells from the sub-menu. If you do it this way, it ought to work for the whole table.

Task 5.12 參考答案

請以必備語庫 5.3 核對答案。

　　現在請嘗試以這些字串寫出一篇文章。請做下面的 Task，可以上面的文章做範本。

Task 5.13

請寫一封電子郵件，解說標示文件中文字的幾種不同方法。（以「反白標示法」為例）

Vocabulary

merge [mɝdʒ] *v.* 合併　　　　　　　**cell** [sɛl] *n.* 儲存格
highlight [ˈhaɪ͵laɪt] *v.* 使……反白；強調　**split** [splɪt] *v.* 分割
cursor [ˈkɝsɚ] *n.* 螢幕上的游標　　　**a dotted line** 虛線

If you want to highlight certain words, or whole sections of text, you can do this in a number of ways. First, highlight the section of the text you want to emphasize by double-clicking on a word, or by left-clicking the mouse and dragging the cursor over the section of text. To make the font bigger, you can either increase the font size in the toolbar at the top of the screen, or go to Format on the menu bar, click Font, and then choose your font size there. If you want to change the color, you should use the color bar on the Drawing tool bar. If you can't find the Drawing toolbar, it could be turned off.

好，現在應該嘗試將學過的字串，套用在自己工作上的電子郵件了。請做下面的 Task，觀摩 Task 5.6 中各行業人士如何回信說明問題的解決方法。

Task 5.14

請根據 Task 5.6，在下面空格中填入正確的名字，完成下列電子郵件。

①

Dear _____,

Thanks for your email and sorry to hear that you are having problems with customs. The documentation may have gotten mixed up in the shipping office. Have you tried calling them to get the correct shipment number? If this doesn't work, let me know and I will contact them myself. Regarding the shipment itself, please check that customs are storing it in cold storage otherwise the goods may perish. Also, make sure you contact the retailer to explain the delay.

Vocabulary

drag [dræg] *v.* 拖；拉   **font** [fɑnt] *n.* 字型

②

Dear _____,

Thanks for your email. I do not see why the new system is so hard to implement. It is used by all sizes of companies all over the region. You just have to make sure that all the figures are kept up-to-date, preferably at the end of each week. If you don't know how to use the system, I can certainly organize some more training. In the meantime, what about if you read the manual again? If every transaction is entered correctly into the chart, then it ought to work perfectly.

Please continue to use it. You will soon get the hang of it.

Task 5.14 參考答案

請閱讀下列摘要，應該不難看出答案。

Email 摘要

1. 瑪莉回信給麥可說明貨品的處置問題。她建議麥可打電話給送貨公司，索取原本的貨單號碼，提醒他確定貨品存放在冷藏庫中以免腐敗，並且要聯絡零售商，解釋貨品遲遲未送到的原因。
2. 喬治回信向吉娜解說新會計系統的問題。他請吉娜沿用該系統，建議她仔細研究使用手冊。他會再安排一場訓練課程。他提醒吉娜，金額資料必須不斷更新，系統才會正常運作。

　　還記得在上一節末，你運用學到的字串來描述自身工作上的問題嗎？現在，試著運用在這節中學到的字串寫封回信，提供問題的解決方法吧！

Vocabulary

get the hang of 熟悉某物的用法；掌握做某事的要領

<語感甦活區> 實用動詞與形容詞

接下來 Task 的重點將是幫你增加關於問題（problem）和解決方法（solution）的兩種字串。請做下面的 Task。

Task 5.15

請研讀下列 word partnerships，然後練習自己造句。

動詞		形容詞		
address	think over	acute	recurring	
appreciate	focus on	awkward	related	
cause	face up to	basic	real	
come across	foresee	central	serious	
come up against	identify	common	tricky	
	ignore	complex	underlying	
confront	overcome	difficult	unexpected	problem
consider	pose	familiar	unforeseen	(+ with sth.)
deal with	present	fundamental	pressing	
define	rectify	main	potential	
detect	run into	minor	outstanding	
encounter	solve			
exaggerate	sort out			
examine	tackle			
explain	think through			

Unit
5

Vocabulary

face up to 面對
pose [poz] *v.* 引起；提出
tackle [ˈtækl] *v.* 應付；解決
awkward [ˈɔkwəd] *adj.* 彆扭的

foresee [forˈsi] *v.* 預知；預料
rectify [ˈrɛktəˌfaɪ] *v.* 修正；改正
think through 想通；想透
pressing [ˈprɛsɪŋ] *adj.* 緊急的；急迫的

- We've **encountered** an **unexpected** problem with the system.
  我們使用這個系統時遭遇到預料之外的問題。

- This is a **difficult** and **recurring** problem which we need to **address**.
  這個棘手又一直發生的問題，是我們需要注意的。

- Let's not **exaggerate** the **potential** problems.
  我們不要誇大了這些潛在的問題。

- This situation **poses** a **tricky** problem.
  這個情況引起了一個棘手的問題。

- We are trying to **think through** the problem with the software.
  我們正試著釐清這個軟體所發生的問題。

📖 **Task 5.16**

請研讀下列 word partnerships，然後練習自己造句。

動詞		形容詞		
adopt	produce	acceptable	partial	
agree on	propose	clever	permanent	
arrive at	provide	concrete	possible	
come up with	put forward	drastic	practical	
demand	reach	early	prompt	
favor	reject	easy	quick	solution
find	require	effective	realistic	(+ to sth.)
hit upon	rule out	feasible	sensible	
look for	work out	immediate	simple	
		lasting	temporary	
		neat	viable	
		obvious	unacceptable	

Vocabulary

hit upon 忽然想起
feasible [ˈfizəbl] *adj.* 可實行的

rule out 拒絕考慮；排除……在外
viable [ˈvaɪəbl] *adj.* 可實行的

參考答案（粗體字為上表中的 word partnerships）

- We need to **come up with** a **neat** solution to this problem.
  我們須想出解決這個問題的好辦法。

- We are **looking for** an **effective** solution, but it's not going to be easy.
  我們正在尋找一個有效的解決之道，但這並不簡單。

- The management does not really **favor** this solution, as it will be too expensive to implement.
  管理階層並不真的喜歡這個方法，因為執行的成本很高。

- We hope you can **reach** an **early** solution that we can move forward.
  我們希望你能先完成一個初步的方案，然後我們可以繼續下一步。

- The R&D team have **hit upon** a really **simple** solution to the problem.
  研發團隊靈光一閃，想到解決這個問題的簡單方法。

　　本章到此結束。離開本章前，請花一點時間做兩件事情。第一件事，重新細讀本章前言中 Task 5.1 的電子郵件，與閱讀本章前寫的電子郵件作比較，看一看自己的學習成果。第二件事，請看看前言中設定的學習目標，是否所有項目都達成了呢？如果還有不甚了解的項目，請重新閱讀相關的 Task 和解說。

　　辛苦了！

Unit 6

回覆 Email 的程序

　　我們在第二章中學過發信和回信的差別。當收到的電子郵件有一個以上的請求或疑問時，最大的困擾可能是不知如何清楚明瞭地答覆。回信中常常會出現以下的錯誤：

1. **內容不完整**，也就是沒有解答來信中的所有疑問。
2. **語意不清**，意即收件者無法在回信中找到所有疑問的解答。

　　出現這兩種錯誤時，雙方得寫更多電子郵件來詢問遺漏的資訊，或花額外的時間在電子郵件中找解答，兩者都會嚴重拖延工作效率。如此浪費時間，實在不專業。本章中，我將教你一套簡單的回信程序，即使遇到又長又複雜的電子郵件，回信時也不會犯下前面提到的錯誤。請先做 Task 6.1，稍後再拿本章教的回信程序來比較，以檢視自己的回信步驟是否需要改進。

Task 6.1

請閱讀這封電子郵件，然後寫一封回信。

Dear You,

Thanks for your email regarding your request for new office furniture. Generally speaking I think it is a good idea and one that would help to improve the company productivity. However, could you please supply me with more information so that I can make a proper decision about it? Could you tell me why you think you need new furniture and what kind of furniture you need? Can you also let me know who will supply the furniture, and what your budget is?

I look forward to hearing from you.

Your department head

　　你可以第208頁「延伸寫作題 6.1」的參考答案與自己的答案作比較。
現在請先做 Task 6.2,寫下你在 Task 6.1 中的回信步驟。你的第一個步驟為
何?第二個步驟又是什麼?以此類推。

📑 **Task 6.2**

上一題中你的回信步驟為何?請寫下來,完成以下句子。見範例①。

① First, I *read the initiating email*　　　　　　　　　　　　　　　　.

② Second, I 　　　　　　　　　　　　　　　　　　　　　　　　.

③ Third, I 　　　　　　　　　　　　　　　　　　　　　　　　.

④ Then, I 　　　　　　　　　　　　　　　　　　　　　　　　.

⑤ I also 　　　　　　　　　　　　　　　　　　　　　　　　.

⑥ Finally, I 　　　　　　　　　　　　　　　　　　　　　　　.

Unit

6

　　我們稍後會回來討論這個 Task。不管你在這 Task 中寫下了什麼,希望
你大致上採取了下列五大步驟:

1. 從來信中找出交代事項
2. 一一編號
3. 找出關鍵字串,畫底線
4. 回信中,按照事項的編號,利用關鍵字串寫答覆
5. 回信中使用參照(referencing)set-phrases,抓住回覆重點

本章接下來將更加詳盡地解說回信步驟，同時給你更多的練習機會。繼續往下看之前，請先看看本章的學習目標：

☐ 充分了解回信的五個步驟。
☐ 收到有多項請求的電子郵件時，能夠快速並清楚地寫出回信。
☐ 回信時，以參照（referencing）的 set-phrases 抓住回覆的重點。
☐ 以其他方式抓住回覆重點。
☐ 充分地練習寫回信。
☐ 學會商務電子郵件中的一些關鍵名詞。

　　本章開始前，我要說明一下，由於本章的教學重點是找出交代事項並以英文回信，所以，**本章接下來所有電子郵件的中文摘要皆移至本章最後**，以達到此項練習的目的。做 Task 時，請勿先看中文摘要，先嘗試自己做，這點非常重要。

掌握來信的訊息

步驟一：找出來信中的交代事項

回信的第一個步驟，就是仔細閱讀來信，找出對方交代的事情。現在請做 Task 6.3。

Task 6.3

請閱讀下面的電子郵件，找出交辦的事項。

Dear Macey,

Thanks for your monthly report, which I have recently received. I have read it carefully. However, there are a few things I am not clear about.

Could you explain why the sales figures for Q2 are so much lower than the target figures? You do not really make this clear. Also, can you tell me in more detail why you have included the costs for the XYZ project in Q2?

Would it be possible for you to save the document as an .rtf file and email it to me again? One more thing, I was wondering if you could have it translated into Chinese by the end of next week. That would be great. Thanks.

Ken

Task 6.3　參考答案

寄件者除了索取更詳細的資訊，還有四個特定請求，其中兩個是索取資訊，兩個是要求處理業務，希望你看得出來。如果看不出來，請重新閱讀電子郵件，尋找請求事項。

Unit
6

第二個步驟是為交代事項編號，以便清楚看出各事項的順序。回信時可能有必要按照編號答覆，因此這個步驟不容忽視。請做下面的 Task。

Task 6.4

按照 Task 6.3 電子郵件中交代事項的順序，在左欄中寫下各事項的編號。見範例。

	Ⓐ	Specific request for information regarding the sales figures for Q2
	Ⓑ	Specific request for action to translate the document
1	Ⓒ	General request for more explanation
	Ⓓ	Specific request for information regarding the costs for the XYZ project
	Ⓔ	Specific request for action to resend the document as an .rtf file

Task 6.4 參考答案

Ⓐ 2　　Ⓑ 5　　Ⓓ 3　　Ⓔ 4

現在來看另一封電子郵件，練習這兩個步驟吧。

Task 6.5

請閱讀下面的電子郵件，找出所有的交代事項，並逐一編號。

Dear Martin,

Thanks for taking the time to finish the presentation slides for me. I really appreciate your efforts. However, in order to make the delivery

---

Vocabulary

slide [slaɪd] *n.* 幻燈片

192

of the presentation go more smoothly, I wonder if you could answer the following questions I have. Could you let me know why you have put slides 47 to 52 in that order? To my mind it would be better to order them in a different way as I cannot clearly see the flow of information. Also, do you think you could explain why you have left out of the presentation all the regional sales figures for Q4? I think it is important to show these. Also, on my laptop, the blue background you have used is very dark. Do you think you could lighten it? Or change the color to make the graphs easier to see from a distance. Finally, I was wondering if you could also produce the revised slide show as a booklet and have it bound cheaply but professionally so that we can impress the board.

Many thanks for your help and hard work on this.

Roger

Task 6.5　參考答案

1. General request for answers
2. Specific request for information regarding the ordering of slides 47-52
3. Specific request for information regarding the omission of Q4 sales figures
4. Specific request for action to change or lighten the background color
5. Specific request for action to produce the slide show as a bound booklet

　　希望你全部答對，如果沒有，請再讀一次電子郵件，看看能否找出所有的交代事項。

Vocabulary

laptop [ˈlæptɑp] *n.* 筆記型電腦　　**booklet** [ˈbʊklɪt] *n.* 小冊子
bind [baɪnd] *v.* 裝訂　　**omission** [oˈmɪʃən] *n.* 省略；遺漏

　　這個步驟比較難，因為有時很難看出哪些是關鍵字串。關鍵字串是文中不可或缺的字串，如果刪除，將導致收件者看不懂電子郵件。關鍵字串通常就是 word partnerships。請做下面的 Task，認識關鍵字串。

Task 6.6

請閱讀下面的電子郵件，找出關鍵字串並畫底線。

Dear Mary,

I am having big problems with the files you sent me on Friday. A lot of the information appear to be missing. I don't know whether this is because you have not included it, or if it got lost in transit, but I would be grateful if you could deal with my questions below. First, could you explain why the figures for Region 4 are missing?

Also, do you think you could let me know why the figures for Region 3 are incomplete: they only cover Q1, Q2, and Q3. Where's Q4? Also, who made the table on page 7? The numbers on the vertical axis are all wrong and will need to be changed. Finally, the captions for all graphs are in the wrong order and will need to be fixed.

Please let me have a new and more complete version of the files by COB today.

Janice

Vocabulary

transit [ˈtrænsɪt] *n.* 傳送；運輸　　**table** [ˈtebl] *n.* 表；表格
axis [ˈæksɪs] *n.* 軸；軸線　　**caption** [ˈkæpʃən] *n.* 標題

📑 **Task 6.6** 參考答案

1. Figures for Region 4
2. Figures for Region 3: no Q4
3. Numbers on the vertical axis in table on page 7
4. Order of captions for all graphs

📑 **Task 6.7**

同樣地，請找出 Task 6.5 電子郵件中的關鍵字串，並畫底線。

📑 **Task 6.7** 參考答案

1. Order of slides 47 to 52
2. Regional sales figures for Q4
3. Background blue
4. Final slide show as a bound booklet

　　現在暫停一下，複習已學過的回信步驟吧。目前為止，我們練習過的三個步驟可以通稱為「掌握訊息」，如下：

　　1. 找出來信中的交代事項
　　2. 逐一編號
　　3. 找出關鍵字串，畫底線

　　往下看下一節，學習如何抓住回信重點之前，請找出最近收到的一封電子郵件再次練習上面的三個步驟。自認已經能掌握這三個步驟的要領時，再進入下一節。

Unit
6

上一節中，我們學過辨認來信重點的三個步驟。本節要學的是抓住回信重點的步驟。相信你還記得，這部分的步驟有兩個：

1. 按照編號順序，使用相同的字串寫回信
2. 使用參照（referencing）set-phrases，抓住回覆重點

現在來練習第四個步驟吧。

✅ 步驟四：按照編號順序，使用關鍵字串寫回信

📑 **Task 6.8**

請閱讀下列電子郵件，這是針對 Task 6.3 電子郵件的回信。注意，梅西在信中按照同樣的編號順序，並使用相同的關鍵字串寫回信。請參考電子郵件下方的表格。

Dear Ken,

Thanks for your email. In reply to your queries, please see below. Regarding the difference between sales and target figures in Q2, this is because three of our sales people left the company and no replacements have been hired yet. If you look at the Appendix on p. 8 of the report, you will see that sales per person are up from last year. In answer to your question about the XYZ project, I've included the costs in Q2 following the guidelines you gave the department last year for costing projects. I hope this is OK.

I am sending the report back to you in .rtf format as you requested, and am arranging to have it translated into Chinese. However, we may need more time for this. I will send it over as soon as it is done.

If you have any questions, please do not hesitate to let me know.

Macey

	Dear Macey		Dear Ken
Request 1	The reason sales figures for Q2 much lower than the target figures	Response 1	Difference between sales and target figures in Q2
Request 2	Costs for the XYZ project	Response 2	I've included the costs
Request 3	Save the document as an .rtf file	Response 3	In .rtf
Request 4	Have it translated into Chinese	Response 4	Am arranging to have it translated into Chinese

Task 6.8　參考答案

希望你一眼便能看出，梅西在回信中按照同樣的編號和關鍵字串作答，由於她抓住回信重點，肯恩輕輕鬆鬆便能找到答覆。另外，注意梅西提供了額外的資訊，如第八頁的表格，也告知肯恩她對交代事項已經做何處理，讓對方知道她沒有把交辦事情忘得一乾二淨。

現在再來做 Task 6.9。別忘了，作答時請先將下方的參考答案遮起來！

Task 6.9

請閱讀下列電子郵件，這是針對 Task 6.6 電子郵件的回信。請在下方的表格中，填入交代事項的編號及其關鍵字串。見範例。

Dear Janice,

I do apologize for the missing information. Please see the following answers to your concerns.

Regarding your question about the missing figures for Region 4, I have attached a file which contains them. I only received these figures from the sales team this morning, which explains why I did not include them with the files I sent last week. Further to your question about the incomplete figures for Region 3, we are having problems getting these from the regional sales team because the sales manager is on leave—his wife is in the hospital having a baby. However, we are working hard to contact him and get the figures. Please be patient with us.

In your email you mentioned that the figures on the vertical axis are wrong. I have checked them again and corrected them. Thank you for pointing out this mistake. Also, I have reordered the captions as you requested.

I hope the files are OK now, and I will let you have the figures for Region 3 as soon as I get them from the field.

Regards,
Mary

Dear Mary		Dear Janice	
Request 1	Missing figures for Region 4	Response 1	*I have attached a file which contains them.*
Request 2	Incomplete figures for Region 3: no Q4	Response 2	Ⓐ
Request 3	Wrong numbers on the vertical axis in the table on page 7	Response 3	Ⓑ
Request 4	Wrong order of captions for all graphs	Response 4	Ⓒ

Task 6.9 參考答案

Ⓐ We are working hard to contact him and get the figures.

Ⓑ I have checked them again and corrected them.

Ⓒ I have reordered the captions.

> 瑪莉在回信中重複使用了關鍵字串，並告知資料不全的原因。

◎ 步驟五：使用參照 set-phrases，抓住回信重點

　　不管上次對方是用傳真、電子郵件、面對面或電話和你聯絡，參照 set-phrases 是用來點出當時提到的事項。除了第二章學過的開場白 set-phrases 外，參照 set-phrases 也能用於電子郵件的開端。這種 set-phrases 可用以使收件者注意到，你將針對對方先前所提出的問題解答，非常實用。

Task 6.10

請再次閱讀 Task 6.9 的電子郵件，找出參照 set-phrases 並畫底線。完成後，先繼續往下看，等一下再回來討論這個 Task。

Task 6.11

請將下列參照 set-phrases 分門別類。

Email 必備語庫 6.1

- ◦ As you requested, ...
- ◦ Regarding the n.p., ...
- ◦ In your email you mentioned that + n. clause ...
- ◦ In your email you mentioned n.p. ...
- ◦ As per your query about n.p., ...
- ◦ Regarding your enquiry about n.p., ...
- ◦ As per your request for n.p., ...

- Regarding your query about n.p., ...
- As you mentioned in your email, ...
- In answer to your question about n.p., ...
- As we discussed on the phone, ...
- It was great to talk to you earlier.
- As per our discussion, ...
- Regarding your suggestion about n.p., ...
- Further to n.p., ...

Task 6.11 參考答案

這個 Task 的分類方法不只一個,所以沒有正確答案,但希望你能夠看出 Regarding、As、In 與其他 set-phrases 的用法。

　　寫電子郵件時,你可能已經在開場白中用過很多這類的 set-phrases,這類 set-phrases 經常被用錯,請做下面的 Task,看看你通常會犯哪些錯誤。

Task 6.12

請更正以下 set-phrases 中的錯誤,見範例 ①。

① Regarding to the XYZ project, ...

→ *Regarding the XYZ project, ...*

② Regarding to your query about the sales targets, ...

→ _____

③ Regarding your enquiry about that we need more resources, ...

→ _____

④ Regards your suggestion about increasing the budget, ...

→ _____

⑤ As for our discussion, ...

→ _____

⑥ Per your request for more data, ...

→ _____

⑦ As you had requested, ...

→ _____

⑧ As you mentioned on your email, ...

→ _____

⑨ As to your query about the client, ...

→ _____

⑩ As we discussion on the phone, ...

→ _____

⑪ In your email you mentioned about project costs.

→ _____

⑫ In answer to your question do we need more resources, ...

→ _____

⑬ It was great talk you earlier.

→ _____

⑭ Further our phonecall, ...

→ _____

請以必備語庫 6.1 核對答案。

> 看到以前一直以為用對的 set-phrases 其實都用錯了，可能感到很詫異吧？建議你回到必備語庫 6.1，在寫錯的 set-phrases 旁畫一顆星，以提醒自己注意。

現在來練習將這些 set-phrases 用在電子郵件中吧。

Task 6.13

以下是艾琳寫的電子郵件，請在空格中填入適當的參照 set-phrases。

①

Dear Luke,

_____Ⓐ_____ for leave next month, I can confirm that your request has been granted. Have a good time.

②

Dear Mary,

_____Ⓑ_____ overtime pay, it is not the custom in Taiwan to pay for overtime. I hope this is clear.

③

Dear Sophia,

_____Ⓒ_____ I'm glad you had a good trip, and I look forward to seeing you at the office again on Monday.

④

Dear Oliver,

Thanks for your email. I have spoken to Donald about your concerns and he agrees with the issues you raised.

_____ D _____ on the phone, he feels that we should move as quickly as possible on this in order not to let our competitive advantage slip away.

_____ E _____ about implementation, he would like to meet you next week to discuss your ideas in more detail. Please call him to arrange a time.

⑤

Dear Judy,

Thanks for your message and for sending the report so quickly. _____ F _____ you would like some help with the implementation of this project. In reply I would like to point out that our resources do not allow us to place another person on this project. _____ G _____ for a larger budget, I'm afraid our budgets for this year have already been frozen. I'm sorry about this, but I hope you understand.

① **Ⓐ** Further to your request

Regarding your request

---

② **Ⓑ** As per your query about

Regarding your enquiry about

Regarding your query about

In answer to your question about

---

③ **Ⓒ** It was great to talk to you earlier.

---

④ **Ⓓ** As per our discussion

As we discussed

Ⓔ Regarding your suggestion

In answer to your question

---

⑤ **Ⓕ** In your email you mentioned that

Ⓖ Further to your request

As per your request

Regarding the request

　　希望你的答案和參考答案相差不遠，如果答案有差異，檢查一下 set-phrases 的結尾，確認是否符合句子。現在正是回去看看 Task 6.10 的時候了，核對畫了底線的 set-phrases 是否正確。如果之前畫錯底線，請再做一次。這次你應該比較能夠找到參照 set-phrases 了吧？

　　當然，將重點逐項列出或編號，以當作回信的格式，也不失為抓住重點的另一種好方法。請看下列回信，並與 Task 6.9 的電子郵件作比較。

【重點逐項列出】

Dear Janice,

I do apologize for the missing information. Please see the following answers to your concerns.

- I have attached a file which contains the missing figures for Region 4. I only received these figures from the sales team this morning, which explains why I did not include them with the files I sent last week.
- We are having problems getting the incomplete figures for Region 3 from the regional sales team because the sales manager is on leave—his wife is in the hospital having a baby. However, we are working hard to contact him and get the figures. Please be patient with us.
- I have checked the figures on the vertical axis again and corrected them. Thank you for pointing out this mistake.
- I have reordered the captions as you requested.

I hope the files are OK now, and I will let you have the figures for Region 3 as soon as I get them from the field.

Regards,
Mary

【將重點編號】

Dear Janice,

I do apologize for the missing information. Please see the following answers to your concerns.

1. I have attached a file which contains the missing figures for Region 4. I only received these figures from the sales team this morning, which explains why I did not include them with the files I sent last week.

2. We are having problems getting the incomplete figures for Region 3 from the regional sales team because the sales manager is on leave—his wife is in the hospital having a baby. However, we are working hard to contact him and get the figures. Please be patient with us.

3. I have checked the figures on the vertical axis again and corrected them. Thank you for pointing out this mistake.

4. I have reordered the captions as you requested.

I hope the files are OK now, and I will let you have the figures for Region 3 as soon as I get them from the field.

Regards,
Mary

注意，這兩種回信寫法，均重複使用了來信（Task 6.6）中的關鍵字串。還有第三種寫法，就是複製來信中的交代事項，貼在回信中，然後逐一答覆。請看下面的範例。

【複製來信中的交代事項】

Dear Janice,

I do apologize for the missing information. Please see the following answers to your concerns.

could you explain why the figures for Region 4 are missing?　I have attached a file which contains them. I only received these figures from the sales team this morning, which explains why I did not include them with the files I sent last week.

why the figures for Region 3 are incomplete: they only cover Q1, Q2, and Q3. Where's Q4?　We are having problems getting these from the regional sales team because the sales manager is on leave—his wife is in the hospital having a baby. However, we are working hard to contact him and get the figures. Please be patient with us.

the table on page 7? The numbers on the vertical axis are all wrong and will need to be changed I have checked them again and corrected them. Thank you for pointing out this mistake.

the captions for all graphs are in the wrong order and will need to be fixed I have reordered the captions.

I hope the files are OK now, and I will let you have the figures for Region 3 as soon as I get them from the field.

Regards,
Mary

通常選用這個方法的原因是方便、節省時間，但是，我認為這個方法最不好，容易產生文法、標點符號問題和錯誤，文意也不清不楚。

至此，我們已經學完回信程序的五步驟，請回想一下並盡量背起來。

一、找出來信中的交代事項
二、逐一編號
三、為關鍵字串畫底線
四、回信時，以相同的編號順序和關鍵字串作答
五、使用參照 set-phrases，抓住回信重點

現在該是做延伸寫作題的時候了，記得本章最前面的 Task 6.1 嗎？看看這次你是否能寫得更好。寫完後，請核對參考答案。

延伸寫作題 6.1

請再次閱讀 Task 6.1 的電子郵件，然後寫封回信。

Dear Department Head,

Thanks for your email and for your interest in my request for new furniture.

Regarding your question about why I think we need the furniture, this is because currently the furniture in the office is very old and doesn't give a good impression to visitors to the office. Also, some of the furniture is broken, and this is unsafe for our staff.

Regarding your query about the kind of furniture, in particular we need new chairs, preferably, chairs with wheels. Our current chairs are very old and have no wheels and they are not convenient for our staff. We also need new tables, as the tables we currently have are too low. Further to your question about who will supply the furniture, I know a very good furniture suppplier company located in Banqiao. I am attaching their catalog for you to see. As per your query about the budget, I think 10,000 NT per staff member will be reasonable. I can negotiate a good discount with this supplier if we order enough items.

I hope this answers all your questions. If you need more information, please do not hesitate to let me know.

Best regards,

　　英文中有可數（countable）、也有不可數（uncountable）名詞，相信這你已經知道。但是要記住哪些名詞可數、哪些不可數往往不容易，更何況中英文對一些名詞可數與否的認知不同。本節將讓你練習改正一般人常用錯的關鍵商業名詞。現在請做 Task 6.14。

Task 6.14

請將下列常見商業名詞分門別類，並填入下表中。

money	information	equipment
dollar	fact	people
suggestion	research	paper
help	advice	newspaper
person	job	news
machinery	work	time
machine	man	feedback
merchandise	input	staff

Countable	Uncountable	Both

Unit 6

請利用下列語庫核對答案。

Email 必備語庫 6.2

Countable	Uncountable	Both
dollar	money	work
suggestion	help	paper
person	machinery	time
machine	merchandise	
fact	information	
job	research	
man	advice	
equipment	input	
people	news	
newspaper	feedback	
	staff	

💡 語庫小叮嚀

- a paper (C) 是「報紙」；paper (U) 是指用來寫字的「紙」。
- a time (C) 是「發生次數」；time (U) 則是「時間」的意思。
- a work (C) 是「作品」；work (U) 是指「工作」。前者在商務英文中極少用到，不用背。
- Staff 這個名詞比較特殊，是指一群員工，不能單指一名員工。例如：We have a staff of 25 in our Taipei office.。如果要單指一名員工，可以用 staff-member 表示，例如：One of our staff-members has resigned.。
- Machinery 是集合稱，指同一系統的機械。
- 很多人分不清 people 和 person 的用法，別擔心，其實很簡單。People 是 person 的不固定複數，請勿使用 persons，這麼寫很怪。名詞有固定複數和不固定複數之分，如：table、tables（固定）；child、children（不固定）。

請看下列電子郵件，找出語庫 6.2 中的名詞並畫底線。注意這些名詞的用法。

①

Dear Martin,

The news from the vendor is that the merchandise is on its way, but has been held up in port by the typhoon. When I know more, I'll let you know.

②

Dear June,

I'm afraid we do not have enough money in our account to pay the staff next month. Please advice.

③

Dear All,

Please make sure you replace the paper in the printer when you finish your work.

④

Dear Vivian,

I have received some information from the client regarding their advertising campaign. The feedback from the research house suggests that the newspaper is not the most effective advertising medium for their product. They would like our help in suggesting alternative media, and want our input at their next strategy meeting on Monday.

Please confirm that this is OK with you.

⑤

Dear Jarvis,

Regarding our conversation about the machinery earlier today, it seems as if three of the machines are broken and will need to be replaced. I have already made a few suggestions to my line manager as to how to avoid this problem, but he refuses to listen to my advice. I have done a lot of research into this topic, and I understand the stress that our procedures put on the equipment.

The following are the facts:

1. There are too many jobs going through the system at the same time.
2. There are not enough people to oversee the work. We need one person per job, not per five jobs as we currently have. Five jobs is too many for one person.

Please have a look at the manufacturer's website and let me know what you think.

Task 6.15 參考答案

答案應不難找出，注意，重點是這些名詞的用法。

終於來到本章結尾。繼續看下一章前，別忘了回到本章的前言看一下學習目標勾出已學會的項目。希望你能全部都打勾！

Email 摘要

Task 6.3

肯恩對梅西的月報告有些疑問，想知道 Q2 的銷售額為何比預期的業績目標低，還有為何梅西在 Q2 中提到 XYZ 企劃案的成本。另外，他希望梅西以 .rtf 檔重寄報告，並將報告翻譯成中文。

Task 6.5

羅傑對馬丁為董事會準備的簡報投影片有些疑問。他覺得第 47 到 52 張投影片的順序不大對，資訊流程不清楚，希望馬丁解釋為何如此排序。還有為何刪除了 Q4 的地方銷售額？投影片的背景好像也太暗，希望馬丁更換或將顏色調淡。此外，他希望投影片可以裝訂成冊，給董事們看。

Task 6.6

珍妮絲收到瑪莉在星期五寄的幾個檔案，但檔案有些問題，例如第四區所有的數據和第三區 Q4 的數據等，很多資料不齊全。第七頁圖表縱軸的數字有問題，圖表的標題順序也有錯。珍妮絲希望今天下班前收到檔案的正確版本。

Task 6.8

梅西回信給肯恩，依交代事項逐一解釋。Q2 的銷售額較低，是因為行銷部人力減少。但是比起去年，每個行銷人員的銷售額度都提高了。梅西在 Q2 附上 XYZ 企劃案的成本，是按照肯恩去年給她的建議所做的。這次她以 .rtf 檔重寄報告，並正在找人將報告翻譯成中文，但請肯恩多給一點時間。

Task 6.9

瑪莉回信給珍妮絲，答覆她在信中的問題。這次她將第四區的完整銷售數據寄出，因為她直到今天早上才收到該區業務部寄來的銷售報告。第三區行銷部經理請假陪妻子生產，因此 Q4 的銷售數據尚不包括該區，請珍妮絲耐心等候。瑪莉已經核對並更正第七頁表格中錯誤的數字，也重新排列所有圖表的標題。她謝謝珍妮絲幫她指出錯誤。

Task 6.13

1. 艾琳批准路克請假的要求。
2. 艾琳告知瑪莉，台灣公司一般沒有支付加班費的制度。
3. 艾琳告訴蘇菲亞，她很期待星期一蘇菲亞收假回到公司崗位。
4. 艾琳回信給奧立弗，告知唐納德聽到他對案子的想法後，表示非常認同，希望他們盡快執行案子。另外他希望在下星期與奧立弗開會，討論案子的執行細節。艾琳請奧立弗打電話給他，安排開會時間。
5. 艾琳回信給茱蒂，答覆她在報告中對企劃案的問題。艾琳說，人事和預算資源不足，無法增加企劃案的人力，茱蒂必須獨力完成該案。

Unit 7

結尾
Set-phrases

這是 Section 1 的最後一章，在這短短的一章中，我們要學的是如何寫電子郵件的結語。用於結語的 set-phrases 不多，因此本章的教學內容可以說簡單易懂。但是如同我在本書中一再強調的，set-phrases 一不小心便會用錯，本章教的一些用法可能和你從前的用法有意想不到的差異。進入本章前，首先請做下面的 Task。

Task 7.1

請閱讀以下電子郵件，找出結語 set-phrases 並畫底線。

①

Dear Mary,

Thanks for taking the time to meet with me yesterday so that I could explain our new product features. I hope you found something to interest you.

Please let me know if you have any questions regarding any of the products, or if you would like to place an order.

Best regards,

②

Dear Edward,

Please could you let me have your figures by 12 today. I need to put them into my report.

I look forward to hearing from you.

Best regards,

③

Dear Joyce,

Thanks for your email. Regarding your questions, please talk to Jane about this, as she has taken over responsibility for the project. I am moving to Hong Kong at the end of the month.

If you have any problems, please let me know and I'll see what I can do to help.

Regards,

④

Dear Jeff,

I am very sorry to hear of your resignation. I know that you are probably glad to be moving on, but we are all going to miss you here at Losers Inc. I wish you all the best in your new job.

⑤

Dear Jim,

Regarding the proposal we're supposed to pitch on Monday, I have been working very hard on it all week, but I still have not received all the data you promised you would send. If I don't receive the data by 5 pm today, I will have to let Darcy know that the pitch will not be going ahead due to lack of cooperation from your department.

I would really appreciate it if you could get a move on!

Vocabulary

pitch [pɪtʃ] *v.* 推銷　　　　**get a move on** 趕快

💡 Email 摘要

1. 寄件者寫信感謝瑪莉給他機會，向她展示公司的新產品。他希望瑪莉會決定向他訂貨。
2. 寄件者請愛德華趕快把數據交給她，她的報告中需要這些數據。
3. 寄件者告知喬依絲，他要移居香港，企劃案不再由他負責，有關企劃的事情可以找珍。
4. 寄件者祝福傑夫的新工作一帆風順，並告訴他原來公司的所有同仁都將懷念他。
5. 寄件者跟吉姆說她很失望，向他索取要用在新客戶行銷活動中的數據，迄今尚未收到。她表示若今天下班前還沒拿到，便要向部門經理達西告狀。

　　我們稍後會再回來討論這些電子郵件，請到時再檢查答案。現在請閱讀以下本章的學習目標。看完本章，你應該達到的學習目標如下：

☐ 正確地使用 if set-phrases 作為結語。

☐ 學會以其他種類的 set-phrases 作為結語。

☐ 糾正以前對含有 wish、hope 和 appreciate 等重要結尾 set-phrases 的錯誤用法。

☐ 充分地練習使用這些 set-phrases。

　　我們這就開始學習吧！

　　結尾 set-phrases 中，最常見的就是包含 if 的句子。雖然這些 if 句的結構和我們在第五章學到的一樣，但其目的和意義卻大相逕庭。此外，這種句型有很多常見的錯誤用法。現在請先看看下列必備語庫。你平常使用的有哪些？

Email 必備語庫 7.1

---

* If you have any problems, ...
* If you have any questions regarding this, please let me know.
* If you require further assistance, please feel free to contact me at any time.
* If you have any further questions or concerns, please feel free to contact us at any time.
* If you have any questions, please do not hesitate to call either ... or ...
* If you have any questions about this, please do not hesitate to contact me.
* You should contact ... if there are any problems.
* If you face any problems, contact me at ...
* If you have any problems, don't hesitate to contact ...

💡 **語庫小叮嚀**

• 注意，如果 if 子句為句子的第一部分，就一定要用逗號；如果為句子的第二部分，則不可用逗號。

• 注意，如果 any 後面接的是可數名詞，該名詞後面務必加 s。

• 注意，如果將 contact 用作名詞，其 chunk 是 be/get/stay/keep in contact with sb.；如果將 contact 用作動詞，其 chunk 是 contact sb.，不須加 with。

• 注意，句子的兩個子句必須互相一致，意即第二個子句中若有任何如 it 或 them 的字，其指的應該是第一個子句中的其中一個名詞。

請改正下列句子中的錯誤，見範例①。做 Task 時不要參考上面的必備語庫。

① If you have any problem, please let me know.

→ *If you have any problems, please let me know.*

② If you have any question regarding this, please let me know.

→ _____

③ If require further assistance, please feel free to contact with me at any time.

→ _____

④ If you have any further questions or concerns, please feel free contact us at any time.

→ _____

⑤ If you have any questions, please do not hesitate call me.

→ _____

⑥ You have any questions about this, please do not hesitate to get in touch me.

→ _____

⑦ You should contact John, if there are any problems.

→ _____

⑧ If you face any problems with this contact me at the office.

→ _____

⑨ If you have any problems, don't hesitate to contact.

→ _____

② If you have any questions regarding this, please let me know.

③ If you require further assistance, please feel free to contact me at any time.

④ If you have any further questions or concerns, please feel free to contact us at any time.

⑤ If you have any questions, please do not hesitate to call me.

⑥ If you have any questions about this, please do not hesitate to get in touch with me.

⑦ You should contact John if there are any problems.

⑧ If you face any problems with this, contact me at the office.

⑨ If you have any problems, don't hesitate to contact me.

現在請嘗試做下面的 Task，做 Task 時不要從上面的語庫找解答，也不要先看參考答案。

📑 Task 7.3

請填空，完成下列 set-phrases，每個空格只能填入一字。

① If you have any _____ regarding this, please send

_____ to me.

② If you have any questions _____ this, please do not hesitate

_____ contact me.

③ If you _____ any problems, contact me.

④ If you _____ further assistance, please _____

free to contact me at any _____.

⑤ If you have _____ problems, don't hesitate to contact

_____.

① questions, them　　　③ have　　　　　　　　⑤ any, me

② about/regarding, to　　④ need/require, feel, time

　　不用說，結尾的 set-phrases 中第二常用的類型就是包含 thanks 的句子。雖然電子郵件可單獨用 Thanks. 作結語，但還有很多其他的 set-phrases 可供選擇。請看看必備語庫 7.2，你經常用到的 set-phrases 有哪些呢？

Email 必備語庫　7.2

- Thank you very much for your help.
- Thanks and sorry for any misunderstanding.
- Many thanks for your understanding on this.
- Thank you very much.
- Thanks in advance.
- Thank you in advance for your time.
- Thanks in advance for your help.
- Thank you again for choosing n.p. …
- Many thanks.
- Thanks for your help.

💡 **語庫小叮嚀**

- Thank 可以當作動詞，也可以當作名詞，使用時要小心。

　　接下來請做 Task，注意 set-phrases 的細節。

📖 Task 7.4

請更正下列句子中的錯誤，然後改寫句子。見範例 ①。

① Thank you on advance for your time.

　→ *Thank you in advance for your time.*

② Thanks in advance to your help.

→ _____

③ Thanks and sorry for a misunderstanding.

→ _____

④ Thank on advance.

→ _____

⑤ Thanks for you help.

→ _____

⑥ Many thanks for your understanding in this.

→ _____

⑦ Many thank.

→ _____

⑧ Thank you again to choose us for your provider.

→ _____

⑨ Thanks you very much.

→ _____

⑩ Thank your very much for you help.

→ _____

🔍 **Task 7.4** 參考答案

現在請利用必備語庫 7.2 核對答案。你一定一個錯誤都沒放過，對不對？

其他種類的 Set-phrases

　　電子郵件的結語除了包含 if 和 thanks 的 set-phrases 外，還有許多其他常見的 set-phrases 可用。我們現在來學一學吧！請看必備語庫 7.3。

Email 必備語庫 7.3

- I look forward to meeting you again soon.
- Have a good day.
- I look forward to your reply.
- Have a good weekend.
- I look forward to working with you on this project.
- Talk to you soon.
- I look forward to hearing from you soon.
- Feel free to contact me at any time.

💡 **語庫小叮嚀**
- 注意 I look forward to 後面一定要接 Ving 或 n.p.。
- 注意 look forward to 前面通常接 I 或 we。

Task 7.5

請重組下列句子，見範例 ①。

① forward / soon / again / I / to / you / look / meeting.

　　→ *I look forward to meeting you again soon.*

② day / have / good / a.

　　→ _____

③ reply / I / forward / look / to / your.

→ _____

④ weekend / have / good / a.

→ _____

⑤ look / working / you / with / this / on / I / forward / project / to.

→ _____

⑥ you / talk / to / soon.

→ _____

⑦ free / feel / contact / at / me / time / any / please / to.

→ _____

Task 7.5 參考答案

請回頭參考必備語庫 7.3。

現在來看一般寫電子郵件時常見的信末頌候語。請看必備語庫 7.4。

Email 必備語庫 7.4

◆ Best regards,	◆ Yours,
◆ Best wishes,	◆ Yours sincerely,
◆ Sincerely,	◆ Yours truly,
◆ Regards,	◆ Yours faithfully,
◆ Sincere best wishes,	◆ Cheers,

💡 語庫小叮嚀

• 你可能還記得以前學寫商業書信時，如果在開端的知照語和稱謂用 Dear
 Sir/Madam，則最後的頌候語要用 Yours faithfully；如果稱謂用收件者的名

字，則最後的頌候語要用 Yours sincerely。這個原則不好記，該寫法較正式，但在現今電子郵件中已愈來愈少見。因此往後你不須再特意去記這個原則了。

一般說來，電子郵件的頌候語較隨意。很多人用一種頌候語行遍天下，不管寫信給誰都一樣，當然也無妨，但建議有時還是改一下比較好。不必擔心頌候語是否恰當或有禮，因為大部分人一天中收到的電子郵件太多，多半對頌候語看都不看一眼。不過必須記住的是，**切勿使用縮寫如 Bst rgds**。不要假設收件者看得懂縮寫，他們的英文不見得比你好，而且會給人懶惰的印象！

＜語感甦活區＞
hope、wish、appreciate

　　在本章的「語感甦活區」中，我們要學結尾 set-phrases 中的三個關鍵字。這三個字常有人覺得很難，也常用錯。但是在學這三個字之前，我們先來看看你平常對這三個字的用法。請做下面兩個練習。

Task 7.6

（一）請分別用 hope、wish、appreciate 造三個句子。句子必須適用於商務電子郵件的結語。

（二）拿出最近自己寫的幾封電子郵件，看看你對這三個字的用法。找出用到這三個字的句子並寫下來。

　　做完這兩個練習後，請研讀下列必備語庫中的結尾 set-phrases。

Email 必備語庫 7.5

* I hope this is clear.
* I hope we can move this along quickly.
* We wish you all a happy ...
* I hope to hear from you soon.
* I hope this is OK.
* I wish you great success in ...

💡 語庫小叮嚀

• 注意 hope 後面接的 n. clause 中永遠不會用 could。

• 注意 wish 後面先接 you 或 you all，然後再接 n.p.。

• 注意 wish 後面接的通常不是 n. clause。

• 注意兩個動詞前面都是接 I 或 we。

Unit
7

✅ Hope 和 wish 的用法

　　你可能還記得以前學 hope 時，課本上說 hope 是指某個可能或真實的事物，wish 則是指某個不可能或虛幻的事物。但事實上這個解釋只是指它們本身的意思，**當這兩個字用在商務電子郵件的結尾 set-phrases 中，並無特殊意義，不過是用來表示祝福而已，沒有意義上的差別**。因此別擔心這兩個字語意上的差異，這在電子郵件中並不重要，只要注意這兩個字的用法即可。請再回頭溫習「語庫小叮嚀」，確定都懂了後，再做下一個 Task。

📑 Task 7.7

請重組下列 set-phrases 或句子，見範例 ①。

① clear / I / this / hope / is.

　→ *I hope this is clear.*

② hope / I / move / along / this / quickly / can / we.

　→

③ year / wish / a / you / new / happy / all / we.

　→

④ in / great / you / new / your / wish / success / job / I.

　→

⑤ OK / this / hope / is / I.

　→

📑 Task 7.7 　參考答案

請回頭參考必備語庫 7.5。

請改正句子中的錯誤，見範例①。

① I wish you have a great holiday.

　→ *I hope you have a great holiday. / I wish you a great holiday.*

② I hope you a Happy New Year.

　→ _____

③ I hope you completion this work soon.

　→ _____

④ I wish you can call me soon.

　→ _____

⑤ I wish you have a good new job.

　→ _____

⑥ I hope receive the fax this afternoon.

　→ _____

📑 **Task 7.8**　參考答案

② I wish you a Happy New Year. / I hope you have a Happy New Year.

③ I hope you can complete this work soon.

④ I hope you can call me soon.

⑤ I wish you great success in your new job.

⑥ I hope I can receive the fax this afternoon.

✅ Appreciate 的用法

好，接下來我們來看看 appreciate 的用法。Appreciate 有三大形式：

appreciate (v.)
appreciative (adj.)
appreciation (n.)

Appreciate 用於 set-phrases 的其中兩種意思為**請求協助**，或者**感謝某人的協助**，第三種意思為**表示了解**，但這種用法比較少見。現在請做下面的 Task。

📑 Task 7.9

請將下列 set-phrases 分門別類，並填入下表中。

- I would appreciate your help with this.
- Your help with this would be much appreciated.
- I really appreciated your help the other day.
- We are always very appreciative of your efforts.
- Please allow me to express my appreciation for your help.
- I appreciate your concerns, but I'm afraid I am not able to help you.
- We very much appreciate your help with this.
- We would very much appreciate it if you could help us.

感謝協助 Thanking for Help

請求協助 Asking for Help

婉拒協助請求 Rejecting a Request for Help

Task 7.9 參考答案

請參照下列必備語庫。

Email 必備語庫 7.6

感謝協助 Thanking for Help
◦ I really appreciated your help the other day.
◦ We are always very appreciative of your efforts.
◦ We very much appreciate your help with this.
◦ Please allow me to express my appreciation for your help.

請求協助 Asking for Help
◦ I would appreciate your help with this.
◦ Your help with this would be much appreciated.
◦ We would very much appreciate it if you could help us.

婉拒協助請求 Rejecting a Request for Help
◦ I appreciate your concerns, but I'm afraid I am not able to help you.

💡 **語庫小叮嚀**

• 注意，請求協助的 set-phrases 都會用 would。

Unit
7

由於 appreciate 的文法比較艱澀，強烈建議你現在就把這些 set-phrases 背起來，同時當然也要注意其中細節，以及正確的用法。

Task 7.10

請以正確形式將 appreciate 填入空格中。

① I would _____ your help with this.

② Your help with this would be much _____.

③ I really _____ your help the other day.

④ We are always very _____ of your efforts.

⑤ I _____ your concerns, but I'm afraid I am not able to help you.

⑥ We very much _____ your help with this.

⑦ We would very much _____ it if you could help us.

Task 7.10　參考答案

請回頭利用必備語庫 7.6 檢查答案。

好，現在是回去看本節 Task 7.6 的時候了，在以前所寫的郵件中，你現在是否已經能找出並改正其中的錯誤了？接下來練習用 set-phrases 寫電子郵件吧。請做下列 Task，做完後再看摘要。

Task 7.11

請用結尾 set-phrases 填空，完成下列電子郵件。

①

Dear Susan,

Please find the attached file for your information. _____Ⓐ_____

②

Dear Mary,

Just to let you know that the negotiation went very well and that we have accepted their offer. They are very excited about the deal and hope that we can implement the project as quickly as possible. _____Ⓑ_____

_____Ⓒ_____

③

Dear Mark,

Regarding our phone calls last week, I apologize for the delay in getting back to you, but it has been really busy at our end. Anyway, regarding your question, the serial number for product XYZ has been changed to 12345ABC. To enter the system, please use the temporary user password below. _____Ⓓ_____ _____Ⓔ_____

④

Dear Alison,

Could you let me know when you will be able to complete the timeline for the new project? The client is asking about our schedule. _____Ⓕ_____

📑 **Task 7.11** 參考答案

① Ⓐ Have a good day.

   Talk to you soon.

   If you have any problems, call me.

② Ⓑ I really appreciated your help the other day.

   Ⓒ I look forward to working with you on this project.

③ Ⓓ I hope this is clear.

I hope this is OK.

Many thanks.

Thanks for your help.

Ⓔ If you have any questions regarding this, please send them to me.

If you have any questions about this, please do not hesitate to contact me.

---

④ Ⓕ If you have any questions regarding this, please let me know.

If you have any questions about this, please do not hesitate to contact me.

Thanks in advance.

Thank you in advance for your time.

Thanks in advance for your help.

💡 Email 摘要

1. 寄件者提醒蘇珊查收附件中的資料。

2. 寄件者告知瑪莉，協商進行地非常順利，而且已接受對方的提議並希望能盡快完成這個案子。

3. 寄件者向馬克致歉，由於忙碌而遲於回覆馬克的問題。他表示 XYZ 產品的序號已改爲 12345ABC，登入系統時，可使用下方的臨時使用者密碼。

4. 寄件者詢問艾莉森何時可完成專案的時間表。

上面只是一些參考答案，你的答案可能大不相同，沒關係，只要選擇大致符合電子郵件文意的 set-phrases 就可以了。舉例來說，上面的郵件中，馬克收到的訊息極度複雜，所以在這裡用 I hope this is clear. 是非常適合的。

好，做最後一個延伸寫作題的時候到了！這是本書的最後一個練習，我建議你這時花一點時間回去看看其他章中教過的 set-phrases，盡量勤加練習，尤其針對那些初次使用時感到最陌生、新的或難的 set-phrases。

請寫一封電子郵件給後勤部，要求儘速核准公司新的辦公室設計。辦公室設計必須在本週之內通過，因為設計師做任何修改需要十天的時間。

參考答案

Dear Judith,

Regarding our meeting last week in which I showed you the designs for the new company office, please find the attached drawings. Please could you let me know your suggested changes to the plan by the end of the week, as the designer needs 10 days to make any changes to the plans. I hope this is clear, and that you like the design.

If you have questions, or would like to make any further changes, please call me.

Many thanks.

♡ Email 摘要

寄件者附上了新的辦公室設計圖。寄件者想要在週末前知道，茱蒂絲建議怎麼改，因為設計師需要十天來改圖。

　　好，Section 1 的最後一章到此告一段落。看下一個 Section 之前，請回到本章前面的 Task 7.1，看看畫上底線的 set-phrases，你現在進步了多少？最後，請到本章前言看一下學習目標，你達成了幾項？希望所有的項目都能打勾。如果還有不甚了解的地方，請重新閱讀相關章節。

Unit
7

Section 2
工具篇：好用句型及例句

Unit 8
發信的開場白

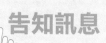

I am writing in response to n.p.
我來信是因為……

例 I am writing in response to a phone call I received from your company this morning. 我來信是因為今天早上我接到你們公司的來電。

I am writing in response to the misleading statements you made about our company on CNN this morning.
我來信是因為今天早上你在 CNN 新聞針對我們公司發表了不實的言論。

I am writing in regard to n.p.
我來信是因為……

例 I am writing in regard to the warehouse supervisor position advertised on your website.
我來信是因為在你們的網站看到倉儲主管職缺的廣告。

I am writing in regard to the problems we've been having with the photocopier we rent from you.
我來信是因為我們向你們租用的影印機一直出問題。

I am writing in connection with n.p.
我來信是關於……

例 I am writing in connection with Judy's request for a promotion.
我來信是關於茱蒂要求升遷的事。

I am writing in connection with the special offer you advertised in the newspaper yesterday.
我來信是關於你們昨天刊登在報紙上的特別優惠。

Vocabulary

warehouse [ˈwɛrˌhaʊs] *n.* 倉庫

supervisor [ˌsupɚˈvaɪzɚ] *n.* 主管；上司

240

I am writing on behalf of n.p.
我是代表……來信。

例 I am writing on behalf of my clients to request immediate payment of the full amount owed.

我代表我的客戶來信，要求立即償付所欠的全額。

I am writing on behalf of Edgeford Insurance to let you know about a few of our exciting new policies.

我代表 Edgeford 保險公司來信，告知你一些我們令人驚喜的新政策。

I am writing to ask about n.p.
我來信詢問有關……

例 I am writing to ask about the possibility of postponing the deadline for the launch of Version 2.3.

我來信詢問關於 2.3 版本延後上市的可能性。

I am writing to ask about the remark you made this morning concerning the closing of the Vancouver office.

我來信詢問有關你今天早上針對關閉溫哥華辦事處所發表的言論。

I am writing to cancel n.p.
我寫信來取消……

例 I am writing to cancel an order I placed earlier today.

我寫信來取消我今天稍早下的訂單。

I am writing to cancel my subscription.

我寫信來取消我的訂閱。

I am writing to request n.p.
我來信是要求……

例 I am writing to request a transfer to the Hong Kong branch.

我來信是想要求轉調到香港的分公司。

I am writing to request punctual payment of the attached invoice.

我來信是要求按照附件的發票準時付款。

I am writing to clear up n.p.
我來信是為了釐清／解決……

例 I am writing to clear up the misunderstanding about the product specs.
我來信是為了釐清關於產品規格的誤會。

I am writing to clear up the problem we had yesterday with our sales person.
我來信是為了解決昨天關於我們業務員的問題。

I am writing to inquire about n.p.
我來信是要詢問……

例 I am writing to inquire about the possibility of a job transfer.
我來信是要詢問調職的可能性。

I am writing to inquire about your new line of snow tires.
我來信是要詢問你們雪胎的新產品線。

I am writing to inform you of n.p.
我寫信來通知你……

例 I am writing to inform you of our decision to close the department.
我寫信來通知你我們結束這個部門的決定。

I am writing to inform you of changes to our billing procedure.
我寫信來通知你我們作帳程序的改變。

I am writing to inform you that + clause.
我寫信來通知你……

例 I am writing to inform you that we will be merging the marketing and sales departments next month.
我寫信來通知你，下個月我們的行銷和業務部門將會合併。

I am writing to inform you that your account has been suspended. If you wish to remain a member, please contact a customer service representative.
我寫信來通知你，你的帳戶已經被凍結。如果你仍然想維持會員身分，請與客服人員聯絡。

I am writing to let you know that + clause.
我寫信來讓你知道……

例 I am writing to let you know that I've resigned and will be leaving Edgeford at the end of this month.

我寫信來讓你知道我已經辭職了，並將於這個月底離開 Edgeford 公司。

I am writing to let you know that we've had some complaints from customers about your attitude.

我寫信來讓你知道，我們已經接到一些關於你的態度的客訴。

I am writing to let you know about n.p.
我寫信來讓你知道……

例 I am writing to let you know about some changes we've made in the way we process payments.

我寫信來讓你知道我們付款方式的一些改變。

I am writing to let you know about our special summer promotion.

我寫信來讓你知道我們夏季的特別促銷活動。

I am writing to tell you that + clause.
我寫信來告訴你……

例 I am writing to tell you that the printer I bought from you last week has not been working well.

我寫信來告訴你，我上星期向你們買的印表機有些問題。

I am writing to tell you that Mary has left the company for personal reasons.

我寫信來告訴你，瑪莉已經因為個人因素離開公司。

I am writing to thank you for n.p./Ving.
我寫信來謝謝你……

例 I am writing to thank you for your help with the project.

我寫信來謝謝你對這個專案的協助。

I am writing to thank you for recommending us to your friend.

我寫信來謝謝你向你的朋友推薦我們。

Please be informed that + clause.
以此通知，……

例 Please be informed that we can't ship your order until we have received payment for your previous order.
以此通知，在收到你上筆訂單的貨款之前我們不會送貨。

Please be informed that our warehouse has been experiencing stock control difficulties and that it may be some time before you receive your order.
以此通知，目前我們的倉儲遇到一些存貨管控的困難，因此你可能需要等待一些時間才能收到訂貨。

Please note that + clause.
請注意……

例 Please note that we will be closed for four days during the Chinese New Year.
請注意，我們在春節期間會停止營業四天。

Please note that I will be out of the office for the rest of the week. I will still be checking email and you can always reach me on my cell phone.
請注意，我這星期接下來都不在辦公室。我仍會檢查我的電子郵件，你打我的手機也都可以找到我。

This is just a quick note to let you know that + clause.
此為一簡短通知，讓你知道……

例 This is just a quick note to let you know that my phone extension has been changed to 522.
此為一簡短通知，讓你知道我的分機已經改為 522。

This is just a quick note to let you know that the order has arrived safely.
此為一簡短通知，讓你知道訂貨已經安全送達。

For your information, + clause.
供你參考，……

例 For your information, I will be leaving the company at the end of the month.
供你參考，我在這個月底會離開公司。

244

For your information, the product shown on page 6 of the catalog has been discontinued and is no longer available.

供你參考，顯示在目錄第六頁的產品已經停產而且不再有貨。

I have been informed that + clause.
我收到通知……

例 I have been informed that we still haven't received your payment.
我收到通知，我們尚未收到你的付款。

I've been informed that starting from next week everyone is required to be in the office before 9:00.
我收到通知，從下星期開始每個人必須九點以前到達辦公室。

We have been asked to V.
我們被要求……

例 We've been asked to check the invoices again. There are some problems with this month's closing.
我們被要求再次檢查發票。這個月的結算有些問題。

We have been asked to give a ten-minute presentation at the shareholders' meeting.
我們被要求在股東會議上簡報十分鐘。

We have just p.p.
我們剛剛……

例 We've just heard that the whole department is being moved to the third floor.
我們剛得知整個部門會搬遷到三樓。

We have just returned from the trade show and have some good news for everybody.
我們剛從貿易展場回來，有些好消息給各位。

Vocabulary

closing [ˈklozɪŋ] *n.*【會計】決算   **shareholder** [ˈʃɛrˌholdə] *n.* 股東

We have received n.p.
我們已經收到……

例 We have received your payment. Many thanks.
我們已經收到你的付款。非常感謝。

We have received no word from you regarding our proposal.
關於我們的提案，我們尚未收到你任何的回應。

n.p. has been received.
……已經收到了。

例 Your request for a credit line increase has been received.
你要提高信用額度的要求已經收到了。

Your payment has been received. Thank you very much.
你的付款已經收到了。非常感謝。

此部分的可能情況較多，請留意 set-phrases 的用法，哪些set-phrases
是接 n.p.；哪些又是接 clause。熟能生巧喔！

We regret having to V.
我們很遺憾必須……

例 We regret having to recall the order, but hope you will understand our position.
我們很遺憾必須撤銷訂單，但希望你可以理解我們的立場。

We regret having to cancel the contract, but we feel we don't have any alternative.
我們很遺憾必須解約，但我們認為我們沒有其他的選擇。

We regret that we have to V.
我們很遺憾我們必須……

例 We regret that we have to suspend construction of the factory.
我們很遺憾我們必須中止廠房的建造。

We regret that we have to cancel the meeting. Let's continue discussing the project by email.
我們很遺憾我們必須取消會面。我們還是繼續用電子郵件來討論這個案子吧。

We regret we cannot V.
我們很遺憾我們無法……

例 We regret we cannot extend our payment terms.
我們很遺憾我們無法延長我們的付款期間。

We regret we cannot offer any discounts at this time.
我們很遺憾我們在現階段無法提供任何折扣。

Vocabulary

alternative [ɔl`tɜnətɪv] *n.* 選擇；替代方案   **term** [tɜm] *n.* 期間；期限

We regret that + clause.
我們很遺憾……

例 We regret that we are no longer able to extend credit to all of our customers.
我們很遺憾無法再延長我們所有客戶的信用期間。

We regret that we have no choice but to lay off three members of your team.
我們很遺憾我們不得不資遣你們團隊的三名組員。

We regret to say that + clause.
我們很遺憾地說……

例 We regret to say that we are not happy with your performance over the last year.
我們很遺憾地說，我們對於你去年的表現並不滿意。

We regret to say that you have violated several of the conditions outlined in the contract.
我們很遺憾地說，你已經違反合約中規範的數條規定。

We regret to tell you that + clause.
我們很遺憾地告訴你……

例 We regret to tell you that we are starting bankruptcy procedures.
我們很遺憾地告訴你，我們已開始辦理破產的程序。

We regret to tell you that we are taking legal action against Edgeford Inc.
我們很遺憾地告訴你，我們正在採取法律行動對抗 Egdeford 公司。

We regret to advise you that + clause.
我們很遺憾地通知你……

例 We regret to advise you that your account has been frozen.
我們很遺憾地通知你，你的帳戶已經被凍結。

We regret to advise you that Tracy has unexpectedly left the company.
我們很遺憾地通知你，崔西出人意料地離開公司了。

Vocabulary

bankruptcy [`bæŋkrəptsɪ] *n.* 破產　　**advise** [əd`vaɪz] *v.*【商業書信】通知

We regret to announce that + clause.
我們很遺憾地宣布……

例 We regret to announce that we have discovered a security issue with the latest version of the software.
我們很遺憾地宣布，我們發現最新版本的軟體有個安全問題。

We regret to announce that as of next month we shall be suspending all business activities.
我們很遺憾地宣布，從下個月開始我們會停止所有的商業活動。

We regret to inform you that + clause.
我們很遺憾地通知你……

例 We regret to inform you that your shipment has been delayed.
我們很遺憾地通知你，你的貨品運送延遲了。

We regret to inform you that the cargo has been impounded.
我們很遺憾地通知你，貨品被扣留了。

I regret to report that + clause.
我很遺憾地報告……

例 I regret to report that we have discovered some irregularities in your reimbursement claim.
我很遺憾地報告，我們在你的退款要求中發現不正常的部分。

I regret to report that sales have continued to decline.
我很遺憾地報告，銷售量持續下滑。

1. 「發布壞消息的 set-phrases」中，動詞都是用 regret 一字，屬於較正式的用法。
2. 每一個 set-phrase 的主詞均可替換，與收件者關係較親近時，可以用 I；關係較遠時可以用 we。

Vocabulary

cargo [ˈkɑrgo] *n.* （裝載的）貨物
claim [klem] *n.* （對應有權利的）要求

reimbursement [ˌriɪmˈbɜsmənt] *n.* 償還；退款
decline [dɪˈklaɪn] *v.* 下降

Did you remember to V?
你記得⋯⋯了嗎？

例 Did you remember to submit a vacation request form?
你記得提出請假單了嗎？

Did you remember to tell Steve about the changes to the project timeline?
你記得告訴史蒂夫關於專案時程的改變了嗎？

Just a quick note to remind you to V.
簡短的通知，提醒你⋯⋯

例 Just a quick note to remind you to set your automatic email reply before you go away.
簡短的通知，提醒你在離開之前設定電子郵件自動回覆系統。

Just a quick note to remind you to give me your report before the end of the month.
簡短的通知，提醒你在月底之前給我你的報告。

Just to remind you that + clause.
只是要提醒你⋯⋯

例 Just to remind you that I will be on leave next week.
只是要提醒你，我下星期休假。

Just to remind you that our phone number has changed. Please see the new number below.
只是要提醒你，我們的電話號碼已經改了，請看下列新號碼。

Vocabulary

submit [səbˋmɪt] *v.* 提出   **timeline** [ˋtaɪmˌlaɪn] *n.* 時程

Just to remind you to V.
只是要提醒你……

例 Just to remind you to turn your phone off before you go into the interview.
只是要提醒你，在面談時先關掉你的手機。

Just to remind you to get all your stuff out of the refrigerator before the end of the week.
只是要提醒你，在每星期結束前把你的東西清出冰箱。

Please remember to V.
請記得……

例 Please remember to turn off the air conditioning before you go home at night.
晚上回家前，請記得關掉空調。

Please remember to tell John I'm still waiting for him to contact me about Judy.
請記得告訴約翰，我還在等他和我聯絡討論茱蒂的事。

Remember to V 是指「記得去（做……）」；而 remember Ving 指的則是「記得（做過……）」。

I am writing to confirm n.p.
我寫信來確認⋯⋯

例 I am writing to confirm your appointment to the position of Marketing Manager. Congratulations, Dick!
我寫信來確認你被任命為新的行銷經理一事。恭喜了，迪克！

I am writing to confirm receipt of your first payment. Your order will ship first thing tomorrow morning.
我寫信來確認你的第一筆款項已收到。你的訂貨明天一大早會送出。

I am writing to confirm that + clause.
我寫信來確認⋯⋯

例 I am writing to confirm that we have received your report before the deadline. Thank you.
我寫信來確認我們在截止日前已收到你的報告，謝謝。

I am writing to confirm that I will be accepting your offer. I'm very much looking forward to working with you.
我寫信來確認我會接受你的提議。我很期待與你共事。

I am pleased to confirm that + clause.
我很高興地和你確認⋯⋯

例 I am pleased to confirm that your application for a credit increase has been approved.
我很高興地和你確認，你提高信用額度的申請已經通過。

I am pleased to confirm that we have received your order. Thank you very much.
我很高興地和你確認，我們已經收到你的訂單。非常謝謝你。

I can confirm that + clause.
我可以確認……

例 I can confirm that we received your report on November 24th.

我可以確認我們在 11 月 24 日收到你的報告。

I can confirm that the trial version will be made available for download on the 15th as scheduled.

我可以確認試用版本會如期在 15 日供大家下載。

I can confirm we have received n.p.
我可以確認我們已經收到……

例 I can confirm we have received your application, but I don't have any other information.

我可以確認我們已經收到你的申請，但我沒有其他資訊。

I can confirm we have received the customer's order, though I'm not sure why we didn't ship it.

我可以確認我們已經收到顧客的訂單，雖然我不確定我們為什麼沒有送貨。

Just a quick note to confirm n.p.
簡短的通知，確認……

例 Just a quick note to confirm receipt of your order.

簡短的通知，確認你的訂單已收到。

Just a quick note to confirm my arrival time. I'll be arriving at 12:20 on Friday morning, which is actually late Thursday night.

簡短的通知，確認我的抵達時間。我會在星期五早上 12 點 20 分抵達，實際上是星期四的深夜。

Just a short note to confirm that + clause.
簡短的通知，確認……

例 Just a short note to confirm that I will be arriving in Sao Paulo on the 13th.

簡短的通知，確認我會在 13 日抵達聖保羅。

Just a short note to confirm that Mr. Katz will not be attending the conference. I apologize for the confusion.

簡短的通知，確認凱茲先生將不會參加會議。我對造成混淆感到抱歉。

Just to confirm that + clause.
只是要確認……

例 Just to confirm that I have booked your hotel as you requested.
只是要確認，我已如你要求地訂了飯店。

Just to confirm that I will need a Mac compatible projector for my presentation.
只是要確認，我在簡報中會需要一個和 Mac 電腦相容的投影機。

This is to confirm that I will be attending n.p.
這封信是來確認我將會參加……

例 This is to confirm that I will be attending the conference, but won't arrive until the second day.
這封信是來確認我將會參加這個會議，但會在第二天才到達。

This is to confirm that I will be attending the meeting and would like to address some of the financing problems that we're facing.
這封信是來確認我將會參加這個會議，而且會提出我們面臨的一些財務問題。

This is to confirm that + clause.
這封信是來確認……

例 This is to confirm that we are already six million dollars over budget.
這封信是來確認我們已經超出預算六百萬元。

This is to confirm that Sophie Ho will succeed Christopher when he retires at the end of the year.
這封信是來確認，在今年底克里斯多福退休後，何蘇菲會接任。

Vocabulary

compatible [kəmˈpætəbl] *adj.* 相容的   projector [prəˈdʒɛktə] *n.* 投影機

This email is to confirm that + clause.
這封信是來確認……

例 This email is to confirm that we have received your registration request.
這封信是來確認，我們已收到你的註冊要求。

This email is to confirm that Eddie Wang worked for our company from 2006 to 2012.
這封信是來確認，王艾迪在 2006 至 2012 年間任職於我們公司。

We hereby confirm n.p.
我們以此確認……

例 We hereby confirm that Mr. Josh Chen is our sole representative for the North East Asia region.
我們以此確認，陳喬栩先生是我們在東北亞地區的唯一代表。

We hereby confirm receipt of payment. Thank you. We do appreciate your business.
我們以此確認付款已收到，謝謝你。感謝你的惠顧。

Confirm 一字後面可接 n.p.，也可接 clause。

Please confirm if + clause.
請確認是否……

例 Please confirm if you would like to go ahead with the project.
請確認你是否想繼續此專案。

Please confirm if you will be attending the conference.
請確認你是否會參加會議。

Please confirm whether + clause.
請確認是否……

例 Please confirm whether you will be making the trip to Seoul or not.
請確認你是否會去首爾。

Please confirm whether you will need a translator for your presentation.
請確認你在簡報時是否需要一位翻譯員。

Please confirm that + clause.
請確認……

例 Please confirm that you will be transferring the entire balance on Monday.
請確認你會在星期一匯足所有餘款。

Please confirm that you have received payment.
請確認你已收到付款。

Please confirm our n.p.
請確認我們的……

例 Please confirm our booking.
請確認我們的預約。

Please confirm our reservation.
請確認我們的訂位。

Please confirm receipt of n.p.
請確認是否收到……

例 Please confirm receipt of our deposit.
請確認是否收到我們的頭款。

Please confirm receipt of our reservation form. Thank you.
請確認是否收到我們的訂位單，謝謝你。

Please give me a call to confirm that + clause.
請給我個電話確認……

例 Please give me a call to confirm that you have received my report.
請給我個電話確認你已收到我的報告。

Please give me a call to confirm that the order has arrived.
請給我個電話確認訂貨已經抵達。

Would you please confirm that + clause?
可否請你確認……？

例 Would you please confirm that you have received our invoice?
可否請你確認你已收到我們的發票？

Would you please confirm that the shipment has arrived?
可否請你確認送貨已經抵達？

If 和 whether 在此的意思相同，都是希望收件者給予「肯定」或「否定」的回答。例如：Please confirm whether/if you can attend the meeting.，收件者應回覆：Yes, I can. 或 No, I cannot.

Vocabulary

deposit [dɪ`pazɪt] *n.* 頭款；押金；訂金

Unit 9

回信的開場白

Thank you for your message.
謝謝你的來信。

例 Thank you for your message. I'm away on business right now, but I'll get back to you when I return.
謝謝你的來信。我現正在出差,回去之後會與你聯絡。

Thank you for your message. Let me look into the situation and get back to you. 謝謝你的來信。讓我了解一下情況再給你回覆。

Thank you for your message about n.p.
謝謝你關於……的來信。

例 Thank you for your message about the new tax law. It really came in handy when we were negotiating the contract.
謝謝你關於新稅法的來信。它在我們協商合約時真的很有用。

Thank you for your message about the new time for the meeting. I was just about to leave the office when I saw it.
謝謝你來信告知新的會議時間。我在剛要離開辦公室的時候看到了。

Thank you for your email.
謝謝你的來信。

例 Thank you for your email. I've forwarded it to the customer service department. In the future you can contact them directly at service@ronko.com.
謝謝你的來信。我已將它轉給客服部,以後你可以來信到 service@ronko.com 直接和他們聯繫。

Vocabulary

on business 出差

look into 調查

get back to sb. 回覆某人(的信件或電話)

come in handy 派上用場;有用

Thank you for your email. I was wondering why I hadn't heard from you. Anyway, glad to hear you're feeling better.

謝謝你的來信。我還在想為什麼沒有你的消息呢。總之，很高興你覺得好一些了。

Thank you for your email about n.p.
謝謝你關於……的來信。

例 Thank you for your email about the new package design. To be honest, I don't like it either.

謝謝你關於新包裝設計的來信。老實說，我也不喜歡。

Thank you for your email about the problems you have been having with the Ronko TX400. We suggest you contact the store where you purchased it.

謝謝你來信提到你對 Ronko TX400 的問題。我們建議你聯絡你購買的商家。

Thanks for your reply.
謝謝你的回覆。

例 Thanks for your reply. I'll pass your suggestions on to John when I see him this afternoon.

謝謝你的回覆。我今天下午遇見約翰時會把你的建議傳達給他。

Thanks for your reply. I understand the problems you've been having with the new software, but I still need a firm launch date.

謝謝你的回覆。我理解你對於新軟體的問題，但我還是需要一個確定的上市日期。

Thank you for sending me n.p.
謝謝你寄給我……

例 Thank you for sending me the report. I'll have a look at it and get back to you if I have any suggestions.

謝謝你寄這份報告給我。我會看看，如果有任何建議會和你聯絡。

Thank you for sending me all those free samples. I'll let you know when we're ready to place an order.

謝謝你寄這些免費的樣品給我。我們準備下訂單時會通知你。

Thank you for sending n.p.
謝謝你寄⋯⋯

例 Thank you for sending the display case so quickly. We got it this morning and it looks great.
謝謝你這麼快就寄來展示箱。我們今天早上收到了，而且看起來很棒。

Thank you for sending Sally's resume to me. It looks like she'd be a great fit for the team.
謝謝你寄莎莉的履歷給我，看起來她應該很適合這個團隊。

Thank you for purchasing n.p.
謝謝你購買⋯⋯

例 Thank you for purchasing a Qwertech mobile phone. We're sure that you'll be happy with your new purchase.
謝謝你購買 Owertech 手機。我們確信你會對你新購買的產品感到滿意。

Thank you for purchasing the best plasma screen television available anywhere, the Ronko TX400. Your warranty has been activated.
謝謝你購買舉世現有最棒的 Ronko TX400 電漿電視。你的產品保固已經生效。

「表達感謝的 set-phrases」是回信的基本用語，不須用到其他種類的 set-phrases 時，都能派上用場。

Vocabulary

purchase [ˋpɝtʃəs] *v.* 購買 *n.* 購買的東西
plasma screen television 電漿電視
warranty [ˋwɔrəntɪ] *n.* 產品保固
activated [ˋæktəˌvetɪd] *adj.* 開始生效的；啓用的

回應壞消息

We're sorry to learn (from sb.) of n.p.
我們很遺憾（從某人）得知關於……

例 We're sorry to learn of the problems you've been having with the elevator. Sam from our technical support team is on his way over to the construction site now.
我們很遺憾你的電梯出了問題。技術支援部門的山姆已經在前往工地的路上。

We're sorry to learn from Jean of your resignation. You'll certainly be missed around here.
我們很遺憾從琴那邊得知你離職的消息。這裡的夥伴一定會想你的。

I'm sorry to learn (from sb.) about n.p.
我很遺憾從（某人）得知關於……

例 I'm sorry to learn from Bill about your decision to find a new supplier.
我很遺憾從比爾那邊得知你要找新供應商的決定。

I'm sorry to learn about your decision not to accept the position in Beijing. Why don't we get together for lunch to talk about it?
我很遺憾得知你不接受北京職位的決定。我們何不一起吃午餐、談談這件事？

I am very sorry to learn (from sb.) that + clause.
我很遺憾（從某人）得知……

例 I am very sorry to learn that Qwertech mobile phone has been giving you problems.
我很遺憾得知你的 Qwertech 手機一直發生問題。

I am very sorry to learn from Queenie that your back has been giving you problems again. I hope you feel better soon.
我很遺憾從昆妮那邊得知你的背又開始不舒服。希望你很快會覺得好一些。

I was sorry to hear about n.p.
我很遺憾得知關於……

例 I was sorry to hear about your recent accident. I wish you a speedy recovery.
我很遺憾得知你最近的意外。希望你會很快康復。

I was sorry to hear about the trouble you had with your Ronko TX400. I suggest that you get in touch with the manufacturer directly.
我很遺憾得知 Ronko TX400 造成你的麻煩。我建議你直接和製造商聯絡。

I was sorry to hear that + clause.
我很遺憾得知……

例 I was sorry to hear that you were disappointed in the service we provided at our hotel in the Maldives. Please accept our offer for a free night's stay the next time you visit.
我很遺憾得知你對於我們在馬爾地夫飯店提供的服務很失望。請接受我們的提議，下次你來時可以免費住宿一晚。

I was sorry to hear that our salesperson was rude to you. We take your satisfaction very seriously and are continuing our investigation of the matter.
我很遺憾得知我們的售貨員對你無禮。我們很在意你是否感到滿意，並持續在調查這個事件。

Sorry 一字是可用在拖延事情時的道歉用語，或者聽到壞消息時的回應。注意，在使用「回應壞消息的 set-phrases」時，that 後面接的是 clause；n.p. 前面則需要介系詞。

Vocabulary

get in touch 聯絡

Apologies for the delay.
對於延遲很抱歉。

例 Apologies for the delay. I have been trying to get more information about your account from the regional office.
對於延遲很抱歉。我一直在設法從區域辦事處取得更多關於你帳戶的資訊。

Apologies for the delay. I've attached the sales figures you asked for.
對於延遲很抱歉。我已附上你要求的銷售數據。

Apologies for the delay in Ving.
延遲……很抱歉。

例 Apologies for the delay in replying to your email. I've been having computer problems, but I think they're all sorted out now.
很抱歉回信給你晚了。我的電腦出了問題，不過我想問題現在都已解決了。

Apologies for the delay in returning your deposit. We'll transfer the money back into your account on the 15th.
延遲歸還你的押金很抱歉。我們會在十五日將錢匯至你的戶頭。

Sorry for the delay in Ving.
延遲……很抱歉。

例 Sorry for the delay in sending out the catalog you requested. You should receive it by the end of the week.
延遲寄出你要的目錄很抱歉。你應該這禮拜之內可收到。

Sorry for the delay in following up on our call the other day. I've had to take a sick day yesterday, so I'm just getting caught up now.
延遲回覆我們幾天前電話中的談話很抱歉。我昨天必須請病假，一直到現在才忙得過來。

Sorry for the delay in responding to n.p.
延遲回覆……很抱歉。

例 Sorry for the delay in responding to your proposal. We've actually been discussing it all week.
延遲對你的提案作出回應很抱歉。事實上，我們一整個星期都在討論它。

Sorry for the delay in responding to your request for a new account.
延遲回覆你開新帳戶的要求很抱歉。

Sorry for not getting back to you earlier, but + clause.
抱歉沒有早些給你回覆，但……

例 Sorry for not getting back to you earlier, but we've been very busy here.
抱歉沒有早些給你回覆，但我們這裡很忙。

Sorry for not getting back to you earlier, but it's taken four days to collect all of the information you requested.
抱歉沒有早些給你回覆，但我們花了四天的時間來收集你要的資料。

Apologies for the delay in getting back to you, but + clause.
回信給你晚了很抱歉，但……

例 Apologies for the delay in getting back to you, but I've been away on leave.
回信給你晚了很抱歉，但我當時在休假。

Apologies for the delay in getting back to you, but we've had problems with our Internet connection.
回信給你晚了很抱歉，但我們的網路連線有問題。

I apologize for the delay in getting back to you, but + clause.
我很抱歉晚回信給你，但……

例 I apologize for the delay in getting back to you, but I've been out of the office.
我很抱歉晚回信給你，但我不在辦公室裡。

I apologize for the delay in getting back to you, but I've just been swamped this week.
我很抱歉晚回信給你，但我這禮拜忙得昏天暗地。

表達祝願

I hope you had a nice weekend.
我希望你的週末過得愉快。

I hope you had a good trip.
我希望你的旅行玩得愉快。

I hope your meeting was successful.
我希望你的會議成功。

I hope you're feeling better now.
我希望你現在覺得好一點了。

I hope this email finds you well.
我希望你身體健康。

I hope you are doing well.
我希望你過得很好。

 雖然這些用語歸類在「回信的開場白」，但視情況也可能出現在 email 的其他部分。

Regarding the n.p., ...
關於……，……

例 Regarding the changes to the contract, I'm still waiting for John to get back to me. 關於合約的變更，我還在等約翰給我回覆。

Regarding the meeting on Thursday, it will be held in downstairs meeting room. 關於星期四的會議，它會在樓下的會議室舉行。

Regarding your question about n.p., ...
關於你對……的問題，……

例 Regarding your question about the meeting on Thursday, it starts at 9:00 and is expected to last for about 2 hours.
關於你對星期四會議的問題，會議會從九點開始，預計開兩小時。

Regarding your question about vacation time, the allowance for your first year is five paid days per year.
關於你對休假時間的問題，你第一年的休假額度是每年五天給薪休假。

As you mentioned in your email, ...
如你在你信中提到的，……

例 As you mentioned in your email, a clean kitchen area is everyone's responsibility.
如你在你信中提到的，乾淨的廚房空間是每個人的責任。

As you mentioned in your email, the problem has been going on for some time.
如你在你信中提到的，這個問題已經持續一段時間了。

Vocabulary

allowance [əˈlauəns] *n.* 限額；允許

In your email you mentioned n.p.
在你的來信中你提到……

例 In your email you mentioned a few of the problems we're having at the
Liaoning plant. What do you suggest we do about them?
在你的來信中你提到我們在遼寧工廠的一些問題。你建議我們怎麼做？

In your email you mentioned Jeremy Chiu as a possible candidate for CFO.
If you'd like, I could ask him if he'd be interested in the position.
在你的來信中你提到邱傑瑞是財務長的可能人選。如果你願意，我可以問他是
否對那職位有興趣。

In your email you mentioned that + clause.
在你的信中你提到……

例 In your email you mentioned that you would like to extend your credit for
another thirty days. I'm afraid that will not be possible.
在你的信中你提到，你想延長信用期間三十天，恐怕那是不可能的。

In your email you mentioned that you'd like to extend the deadline, but
didn't say till when. Can you let me know how much additional time you
will need?
在你的信中提到，你想延長截止期限，但沒說明延至何時。你能讓我知道你需
要多少額外的時間嗎？

In answer to your question (about n.p.), ...
回答你（關於……）的問題，……

例 In answer to your question about the warranty, I'm afraid that it doesn't
cover rain damage.
回答你關於產品保固的問題，恐怕那不包含雨淋的損壞。

In answer to your question about the delivery date, you can expect the
shipment to arrive in three days.
回答你關於送貨日期的問題，你可以預計送貨在三日後抵達。

Vocabulary

plant [plænt] *n.* 工廠　　　　　　　　candidate [ˋkændədet] *n.* 候選人；候補者

As we discussed on the phone, ...

如我們在電話上討論的，……

例 As we discussed on the phone, Mike will be taking over my job, while I'm in Osaka. 如我們在電話上討論的，我在大阪的時候，麥可會接手我的工作。

As we discussed on the phone, Judy will be giving the presentation this time. 如我們在電話上討論的，這次茱蒂會做簡報。

Regarding your suggestion about n.p., ...

關於你對……的建議，……

例 Regarding your suggestion about the online ads, I've already passed it on to Daphne and will let you know what she says. 關於你對線上廣告的建議，我已經轉給戴芬妮，會再讓你知道她的說法。

Regarding your suggestion about the specs, we've decided not to make any changes. Actually, I agree with you, but changes are just too expensive at this stage. 關於你對於規格的建議，我們決定不做任何改變。事實上，我同意你的意見，但在現階段改變的成本太高。

As you requested, ...

如你所要求，……

例 As you requested, here is our Asian currency forecast for the coming three months. 如你所要求，這是未來三個月的亞洲匯率預測。

As you requested, I've attached the marketing copy for the spring campaign. The images will be ready before the end of the week. 如你所要求，我已附上春季活動的行銷文案，圖像部分在這禮拜之內會完成。

It was great talking with you earlier.

先前和你談話很開心。

例 It was great talking with you earlier. I've attached a PDF with our catalog and price list. I'm looking forward to working together. 先前和你談話很開心。我已附上目錄和價目表的 PDF 檔。我很期待與你共事。

It was great talking with you earlier. I hope you enjoy your new job. 先前和你談話很開心。我希望你喜歡你的新工作。

Unit 10

表達請求

Can you V?
你可以……嗎？

例 Can you send me Karen's email address? I can't seem to find her card.
你可以寄給我凱倫的 email 嗎？我好像找不到她的名片。

We have a problem with the Uniponz account. Can you call me as soon as you get this?
我們對 Uniponz 的帳戶有問題，你收到這封信後可以盡快打個電話給我嗎？

Could you V?
你可以……嗎？

例 You're heading over to accounting this afternoon, right? Could you ask Mary to call me if you see her?
你今天下午會去會計部，對吧？如果你看見瑪莉，可以請她打個電話給我嗎？

Could you let me know when you'll be arriving?
你可以讓我知道你何時會抵達嗎？

Do you think you could V?
你覺得你可以……嗎？

例 Do you think you could let me have a copy of the report?
你覺得你可以給我一份這個報告嗎？

Do you think you could arrange for an engineer to come and talk to the sales people about the design?
你覺得你可以安排一位工程人員過來向業務人員說明這個設計嗎？

Vocabulary

card [kɑrd] *n.* 名片

272

Do you think you could possibly V?

你覺得你有可能……嗎？

例 Do you think you could possibly postpone the CEO's visit until next month?
你覺得你有可能把總裁的來訪延至下個月嗎？

Do you think you could possibly have it ready by March 10?
你覺得你有可能在 3 月 10 日前完成它嗎？

Would it be possible for you to V?

你有沒有可能……？

例 Would it be possible for you to let me access your customer service database? I'd like to know exactly who has been experiencing problems and when.
你有沒有可能讓我使用你們的客服資料庫？我想精確地知道哪些人在何時遭遇過問題。

It sounds like a good idea, but would it be possible for you to submit a formal proposal?
聽起來是個好主意，但你有沒有可能提出一個正式的提案？

I wonder if you could V.

我在想你是否可以……

例 I wonder if you could send me the minutes from the final negotiation. I'd like to compare what was said with what was written in the contract.
我在想你是否可以寄給我最後談判的記錄。我想比較一下口頭談的和合約上的記載。

Actually, we're ahead of schedule. I wonder if you could have the packaging ready two weeks ahead of time.
事實上，我們超前預計的時程。我在想你是否可以提早兩個星期完成包裝。

Vocabulary

postpone [post`pon] *v.* 使延期；延遲

minutes [`mɪnɪts] *n.*【複數型】會議記錄

access [`æksɛs] *v.*（取用）資料；使用

ahead of schedule 比預計的時間早

I was wondering if you could V.
我在想你是否可以……

例 I was wondering if you could come to New York so that we can meet face to face about this.
我在想你能否來紐約一趟，我們可以面對面地談談。

I'm going to be on vacation starting from next week, so I was wondering if you could finish the report for me.
我下個星期開始休假，所以我在想你是否可以幫我完成這份報告。

Will you do it?
你要做嗎？

例 I can never get the numbers to come out right. Will you do it?
我永遠沒辦法算出正確的數字。你要算嗎？

I won't have time to write the report. Will you do it?
我沒有時間寫這份報告。你要寫嗎？

Will you help me with this?
你可以幫我處理這個嗎？

例 I'm having problems accessing the database. Will you help me with this?
我在使用資料庫時有問題。你可以幫我處理嗎？

Will you help me with this? I don't think I'll be able to finish it before the deadline without your help.
你可以幫我處理這個嗎？我想沒有你的幫忙我沒辦法在截止日前完成。

Vocabulary

database [ˋdetəˌbes] *n.* 資料庫　　　**deadline** [ˋdɛdˌlaɪn] *n.* 截止期限

Can you tell me wh-clause?
你可以告訴我……嗎？

例 Can you tell me who is responsible for new customer accounts?
你可以告訴我誰負責新的顧客帳戶嗎？

Can you tell me when the new office furniture will be arriving?
你可以告訴我新的辦公設備何時會抵達嗎？

Could you explain wh-clause?
你可以解釋……嗎？

例 Could you explain exactly what you'd like me to do with the report?
你可以解釋清楚你要我怎麼做這個報告嗎？

Could you explain why everyone's passwords are taped to the front of their monitors?
你可以解釋一下為什麼每個人的密碼都貼在螢幕前嗎？

I wonder if you could explain wh-clause.
我在想你是否可以解釋……

例 I wonder if you could explain why the shipment was delayed.
我在想你是否可以解釋送貨為何延遲。

I wonder if you could explain how to fill out the customs form again. I always mess it up.
我在想你是否可以再次解釋如何填寫海關表單。我總是會搞砸。

Vocabulary

fill out 填寫（表格、申請書等）　　　**mess up** 使……混亂；把……搞砸

Would it be possible for you to let me know wh-clause?
你有可能讓我知道⋯⋯嗎？

例 Would it be possible for you to let me know how you arrived at your estimate? 你有可能讓我知道你如何得出這個預估值嗎？

Would it be possible for you to let me know who your current packaging supplier is? 你有可能讓我知道你們目前的包裝廠商是誰嗎？

Do you think you could possibly tell me wh-clause?
你覺得你可以告訴我⋯⋯嗎？

例 Do you think you could possibly tell me why my luggage is in Istanbul? 你覺得你可以告訴我為何我的行李會在伊斯坦堡嗎？

Do you think you could possibly tell me which company Jamie is working for now? 你覺得你可以告訴我傑米現在在哪家公司工作嗎？

Do you think you could let me know wh-clause?
你覺得你可以讓我知道⋯⋯嗎？

例 Do you think you could let me know when my order will be ready? 你覺得你可以讓我知道我的訂貨何時會準備好嗎？

Do you think you could let me know how many people will be staying for lunch after the meeting? 你覺得你可以讓我知道有多少人會在會議之後留下來用午餐嗎？

I was wondering if you could let me know wh-clause.
我在想你是否可以讓我知道⋯⋯

例 I was wondering if you could let me know when you'd like to meet again. 我在想你是否可以讓我知道你什麼時候要再碰面。

I was wondering if you could let me know who was responsible for leaving the lights on all night. 我在想你是否可以讓我知道誰要為燈整夜沒關負責。

Vocabulary

estimate [ˈɛstəmɪt] *n.* 估算；估計   **supplier** [səˈplaɪə] *n.* 供應商

I'll V.
我會……

例 I'll give you a call immediately after the meeting.
我在會議結束後會立刻打電話給你。

I'll do my best to make sure the meeting goes smoothly.
我會盡全力確保會議進行順利。

I'll see what I can do.
我看看我能做些什麼。

例 I'm not sure if I can help, but I'll see what I can do.
我不確定我能否幫上忙,但我看看我能做些什麼。

Leave it with me. I'll take a look and see what I can do.
把它交給我吧,我看一下並想想我能做些什麼。

I'll be back on Thursday.
我星期四會回來。

例 I'm going to Hong Kong, but I'll be back on Thursday.
我要去香港,但會在星期四回來。

I'll be back on Thursday and we can talk about it then.
我會在星期四回來,我們可以到時再談。

I'll be in touch.
我會和你聯絡。

例 Don't worry. I'll be in touch as soon as it's ready.
不用擔心,一準備好時我就會和你們聯絡。

I'll be in touch later this week.
這個星期晚些時候我會和你們聯絡。

I'll give you a call.
我會打電話給你。

例 I'll give you a call after I've spoken to the client and let you know what they say.
我和客戶談過之後會給你個電話,讓你知道他們怎麼說。

Don't bother meeting me at the airport. I'll give you a call after I arrive at the hotel.
不用麻煩到機場來碰面,我到達飯店之後會給你個電話。

I'll drop you a line.
我會寫封信給你。

例 I'll drop you a line after I've spoken to Mike and let you know what he said.
我和麥可談過之後會寫封信給你,讓你知道他說了什麼。

I'll drop you a line after the conference and let you know if there are any new developments.
在會議之後我會寫封信給你,讓你知道有無任何新發展。

I'll send it tomorrow.
我明天就會寄出去。

例 I'm almost done with the prototype. I'll send it tomorrow.
我已經快將原型完成了,我明天就會寄出去。

We're still working out a few last details of the proposal. I'll send it tomorrow.
我們還在處理提案的最後一些細節,我明天就會寄出去。

I'll get back to you as soon as I can.
我會盡快給你回覆。

例 I know it's taking a long time, but please be patient. I'll get back to you as soon as I can.
我知道已經過了很長的時間,但請耐心等候。我會盡快給你回覆。

Let me talk to John about it first and then I'll get back to you as soon as I can.
讓我先和約翰談談,然後我會盡快給你回覆。

Unit 11
安排會面

提議會面

Can you make n.p.?

……你可以嗎？

例 Can you make next Monday at 9:00?

下星期一 9 點你可以嗎？

Can you make our noon staff meeting on Wednesday?

星期三中午的職員會議你能參加嗎？

Can you manage n.p.?

……你可以嗎？

例 Can you manage Thursday morning?

星期四早上你可以嗎？

Can you manage 3:00 on Wednesday the 24th?

24 日，星期三 3 點，你可以嗎？

Is n.p. OK for you?

……你可以嗎？

例 Is 4:30 OK for you?

4:30 你可以嗎？

Is early next week OK for you?

下星期初你可以嗎？

Would n.p. be OK for you?

……你可以嗎？

例 Would Friday be OK for you?

星期五你可以嗎？

Would 8:00 pm be OK for you?

晚上 8:00 你可以嗎？

How is n.p. for you?
……你覺得如何？

例 How is Monday for you? 星期一你覺得如何？

How is next week for you? 下星期你覺得如何？

How about n.p.?
……如何？

例 How about Starbucks? 星巴克如何？

How about the Irish Pub at the Hyatt? 君悅酒店的愛爾蘭酒吧如何？

How about Ving?
……如何？

例 How about talking about it over some lunch?
一起吃午餐討論這件事如何？

How about meeting sometime next week?
下星期找個時間碰面如何？

What about n.p.?
……如何？

例 What about the Intercontinental? They're having a French food festival at the moment. Intercontinental 酒店如何？他們現在有法國美食節的活動。

What about Tuesday the 19th?
19 日的那個星期二如何？

What about Ving?
……如何？

例 What about taking them to the opera?
帶他們去看歌劇如何？

What about having dinner together on Thursday?
星期四一起用晚餐如何？

Let's V.

一起……

例 Let's do lunch. My treat.
一起吃午餐，我請客。

Let's have coffee together and see if we can work something out.
我們一起喝咖啡，看看我們能否把事情解決。

We could V.

我們可以……

例 We could have a coffee together during the conference and discuss it more.
我們可以在開會時喝個咖啡，進一步討論這件事。

We could take them to the Hilton for cocktails and then have dinner at Romero's.
我們可以帶他們到希爾頓飯店喝個雞尾酒，然後到 Romero's 用晚餐。

Why don't we V?

我們何不……？

例 Why don't we meet for a coffee next week?
我們下星期何不碰面喝個咖啡？

Why don't we set up a meeting for later in the month? How about the 25th?
這個月稍晚我們何不碰個面？ 25 日如何？

I'd like to suggest n.p.

我想要提議……

例 I'd like to suggest the Four Seasons. It's a little more expensive, but it's also a lot more convenient.

我想要提議四季飯店。它稍微有點貴，但方便許多。

I'd like to suggest no later than Friday.

我建議不要晚於星期五。

I'd like to suggest that + clause.

我想要提議……

例 I'd like to suggest that we invite Frank to join the meeting.

我提議我們邀請法蘭克來參加會議。

The Metropolitan is going to be packed, so I'd like to suggest that we try the Ritz instead.

大都會飯店一定會人滿爲患，所以我提議我們改在麗池飯店。

When would be convenient for you?

什麼時候對你比較方便呢？

例 I'd like to discuss this face to face. When would be convenient for you?

我想要面對面討論這件事，什麼時候對你比較方便呢？

Let's get together and bounce some ideas around. When would be convenient for you?

讓我們見個面並且激盪一些想法，什麼時候對你比較方便呢？

Vocabulary

bounce around 從一個人傳給另一人

n.p. is no good for me.
……我不行。

例 Unfortunately, Monday is no good for me.
真不巧，星期一我不行。

Sorry, 6:00 is no good for me. I've got to pick my daughter up from her ballet lesson.
抱歉，六點我不行。我女兒上完芭蕾舞課我要去接她。

n.p. is not OK for me.
……我不行。

例 The Hyatt is not OK for me. Isn't there somewhere closer that we could meet?
君悅酒店我不行。沒有近一點的地方可以碰面嗎？

The auditors will be here from the 7th, so next week isn't OK for me.
稽查員從七日起會待在這邊，所以下星期我不行。

I can't make it then.
我那時候不行。

例 I'll be away on business, so I can't make it then.
我將會在出差，所以那時候我不行。

I'm in Taichung that day, so I can't make it then.
那天我人在台中，所以那時候我不行。

Vocabulary

pick up 接（人）

auditor [ˈɔdɪtə] *n.* 查帳員；稽查員

I can't manage it then.
那時候我不行。

例 I'm on leave next week, so I can't manage it then.
下星期我休假，所以那時候我不行。

Next week? I can't manage it then. The week after is better for me.
下星期？我那時候不行。再下個星期對我比較方便。

My schedule is really tight (+ time).
（時間）我的行程真的很緊湊。

例 My schedule is really tight next week.
下星期我的行程真的很緊湊。

My schedule is really tight. Can we have a conference call instead of a meeting?
我的行程真的很緊湊，我們可以用電話會議來取代會面嗎？

正式

I'm afraid I have to V.
我恐怕必須……

例 I'm afraid I have to be in Singapore next week.
下星期我恐怕會在新加坡。

I'm afraid I have to meet another client at 6:00. How about tomorrow morning?
六點我恐怕必須見另一個客戶，明天早上如何？

I'm afraid I'm going to be Ving.
我恐怕必須……

例 I'm afraid I'm going to be meeting with the marketing team at that time.
那時候我恐怕必須和行銷團隊開會。

I'm afraid I'm going to be conducting a factory visit then.
那時候我恐怕得帶領參觀廠房。

I'm afraid I'm going to be adj.
我恐怕會……

例 I'm afraid I'm going to be unavailable then.
我那時候恐怕會不在。

I'm afraid I'm going to be on leave next week.
我下星期恐怕會休假。

I'm afraid I'm tied up then.
我那時候恐怕有事。

例 Next Monday is the start of the new quarter. I'm afraid I'm tied up then.
下星期一是新一季的開始，我那時候恐怕會沒空。

I've got lots of meetings on Thursday, so I'm afraid I'm tied up then.
星期四我有很多會要開，所以我那時候恐怕會沒空。

I'm afraid that's not possible.
恐怕那是不可能的。

例 I'm away on business all next week, so I'm afraid that's not possible.
下一整個星期我都在出差，所以恐怕那是不可能的。

Regarding your request for an appointment with the senior vice president, I'm afraid that's not possible.
關於你要和資深副總會面的要求，恐怕那是不可能的。

Vocabulary

quarter [ˈkwɔrtə] *n.* 一季；季度　　　　**tied up** 忙得無法脫身的

286

Unit
11

非正式

n.p. is good for me.
……我可以。

例 Monday is good for me. 星期一我可以。

　　3:00 is good for me. 三點我可以。

n.p. is OK with me.
……我可以。

例 Yes, nine in the morning is OK with me. Even eight wouldn't be too early.
　　是的，早上九點我可以。就算是八點也不會太早。

　　The hotel bar is OK with me. I'm not picky.
　　飯店酒吧我可以，我不會很挑剔。

I'll be free at n.p.
……我有空。

例 I'll be free at 4:30 in the afternoon. 下午 4:30 我有空。

　　I'll be free at the beginning of next week. 下星期初我有空。

正式

That would be fine. 那應該沒問題。

That's fine. 沒問題。

That's good for me. 我沒問題。

Yes, I can make/manage it. 是的，我可以。

sb. be going to V.
……將會……

例 I'm going to visit our factory in Suzhou next week.
下星期我將會去參觀我們在蘇州的廠房。

Tom's going to study the report over the weekend.
湯姆週末要研究這份報告。

sb. be Ving.
……會……

例 I'm having lunch with Julian.
我會和朱利安一起吃午餐。

Karen is leading the discussion at the team meeting this Friday.
凱倫會在本週五的小組會議帶領討論。

I'll be waiting ...
我會等……

例 I'll be waiting in the lobby for you.
我會在大廳等你。

Don't forget to contact me after the meeting. I'll be waiting for your call.
別忘了會議結束後和我聯絡，我會等你的電話。

I'll be arriving ...
我會抵達……

例 I'll be arriving at the airport at five in the morning.
我會在早上五點抵達機場。

I'll be arriving by train on Monday afternoon.
我會在星期一下午搭火車抵達。

I'll be leaving ...
我會離開……

例 I'll be leaving immediately after the meeting.
會議結束後我會立刻離開。

I'll be leaving on Friday.
我會在星期五離開。

I'll have to V.
我必須……

例 I'll have to leave no later than 8:00.
我最晚必須在八點以前離開。

I'll have to arrange for someone to take you to the airport.
我必須安排個人帶你到機場。

I've got a meeting ...
我有個會議……

例 I'll have to stop by on Friday because I've got a meeting on Thursday.
因為我星期四有個會議，我星期五會順道拜訪。

I've got a meeting with Jason at his office first thing Monday morning.
我星期一一大早和傑森在他的辦公室開會。

I've got an appointment ...
我有約……

例 I've got an appointment at 8:30, so lunch should be no problem. See you at 1:30?
我八點半有約，所以午餐應該沒問題。一點半見了囉？

I've got appointments on Tuesday, Thursday, and Friday afternoon.
我星期二、星期四和星期五下午都有約了。

I need to change/cancel/postpone n.p.
我需要更改 / 取消 / 延後……

例 I need to change the venue.
我需要更改地點。

I need to cancel the conference.
我需要取消會議。

I need to postpone the meeting until next week.
我需要將會議延後至下星期。

I'm afraid I have to V.
我恐怕必須……

例 I'm afraid I have to meet a new client at that time.
我那時候恐怕必須見個新客戶。

I'm afraid I have to leave the office on time Friday night.
我星期五晚上恐怕得準時離開辦公室。

I'm afraid we're going to have to cancel n.p.
我們恐怕得取消……

例 I'm afraid we're going to have to cancel the meeting.
我們恐怕得取消會議。

I'm afraid we're going to have to cancel our appointment.
我們恐怕得取消我們的約定。

Vocabulary

venue [ˈvɛnju] *n.* 會場；舉辦地點

I'm afraid we're going to have to postpone n.p.
我們恐怕得將……延後。

例 I'm afraid we're going to have to postpone the launch date.
我們恐怕得將上市日期延後。

I'm afraid we're going to have to postpone the entire project until next year.
我們恐怕得將整個專案延後至明年。

I'm going to have to V.
我必須……

例 I'm going to have to ask Chester to turn control of the project over to Selena.
我必須要契斯特把專案的掌控權移轉給席琳娜。

I'm sorry, but I'm going to have to reschedule our meeting. Will you have time tomorrow afternoon?
我很抱歉，但我必須重新安排會議的時間。明天下午你有空嗎？

Would it be OK if we postponed it till ...?
如果我們把它延後至……，可以嗎？

例 Would it be OK if we postponed it till after the trade show?
如果我們把它延至商展之後，可以嗎？

Would it be OK if we postponed it till next quarter?
如果我們把它延至下一季，可以嗎？

Vocabulary

turn over 移交
trade show 商展；貿易展

reschedule [riˋskɛdʒul] v. 重新安排時間

Unit 12
急件

As the client has requested n.p. by ..., please do this as soon as you can.

客戶要求在……前……，所以請盡快處理。

例 As the client has requested an answer by the end of the week, please do this as soon as you can.

客戶要求在本週之內回覆，所以請盡快處理。

As the client has requested the report by the end of the month, please do this as soon as you can.

客戶要求在月底前回報，所以請盡快處理。

If we have not received n.p. by ..., we will have no choice but to V.

假如我們在……前沒有收到……，我們就只好……。

例 If we have not received payment by the end of the week, we will have no choice but to suspend the service.

假如我們在本週結束前沒有收到帳款，我們就只好中止服務了。

If we have not received your final answer by Friday, we will have no choice but to cancel the project.

假如我們在週五前沒有收到你的最終答覆，我們就只好把案子撤銷了。

Unless we receive n.p. by ..., we will be forced to V.

除非我們在……前收到……，否則我們將被迫……。

例 Unless we receive payment by the first of the month, we will be forced to take legal action against you.

除非我們在月初前收到帳款，否則我們將被迫對你們採取法律行動。

Unless we receive the goods by the end of the month, we will be forced to take further action to recover our money.

除非我們在月底前收到貨，否則我們將被迫採取進一步的行動來討回款項。

We need to have n.p. by ...
我們要在……前收到……。

例 We need to have the report by the 23rd.
我們要在二十三日前收到報告。

We need to have your reply by the end of the day.
我們要在今天結束前收到回覆。

It is essential that we receive it before ...
我們一定要在……前收到。

例 It is essential that we receive it before the holiday.
我們一定要在假日前收到。

It is essential that we receive it before you go away on vacation.
我們一定要在你休假前收到。

It is vital that we receive it before ...
我們非得在……前收到不可。

例 It is vital that we receive it before they do.
我們非得在他們行動前收到不可。

It is vital that we receive it before production starts.
我們非得在開始生產前收到不可。

It is critical that we receive it before ...
我們必須在……前收到才行。

例 It is critical that we receive it before close of business Monday.
我們必須在星期一下班前收到才行。

It is critical that we receive it before the end of the month.
我們必須在月底前收到才行。

Vocabulary

essential [ɪˈsɛnʃəl] *adj.* 必要的   vital [ˈvaɪt!] *adj.* 必不可少的

critical [ˈkrɪtɪk!] *adj.* 緊缺而必須的

..., otherwise we will be forced to V.
……，否則我們將被迫……

例 We must complete the transfer as soon as possible, otherwise we will be forced to cancel the payment.
我們必須盡快完成轉帳，否則我們將被迫取消付款。

We must receive the product as soon as possible, otherwise we will be forced to withhold payment.
我們必須盡快收到產品，否則我們將被迫暫停付款。

..., otherwise we will be unable to V.
……，否則我們就沒辦法……

例 We need the figures really soon, otherwise we will be unable to complete the report. 我們馬上就需要這些數字，否則我們就沒辦法完成報告了。

You must send us the specs as soon as you can, otherwise we will be unable to make the changes you requested.
你一定要盡快把規格寄給我們，否則我們就沒辦法照你們的要求來修改了。

..., so we need n.p. by ...
……，所以我們在……前需要……

例 Our competitors are catching up, so we need a new strategy by the end of the year.
我們的競爭對手正在迎頭趕上，所以我們在年底前需要一套新的策略。

We are running out of this item, so we need a new order by July.
我們快用完這個品項了，所以我們在七月前需要訂新貨。

Vocabulary

withhold [wɪð`hold] *v.* 不給；保留

..., so we hope that you can V at the soonest possible time.
……，所以我們希望你們盡量能在最短的時間內……

例 We are allocating resources for this project, so we hope that you can confirm at the soonest possible time.
我們正在為這個案子分配資源，所以希望你們盡量能在最短的時間內確認。

We need to prepare our team, so we hope that you can give us the go-ahead at the soonest possible time.
我們需要為團隊作準備，所以希望你們盡量能在最短的時間內下達指令。

We hope that you can send it at the earliest possible time.
我們希望你們能盡早把它寄來。

There is not much time left.
所剩時間不多了。

Time is running out.
時間快不夠了。

We do not have much lead time for this.
我們沒有多少時間可以做這件事了。

Please give this matter your urgent attention.
請趕緊處理這件事。

I need this done as soon as possible.
我要這件事盡快完成。

I would be grateful if you could make this top priority.
假如你能把這件事列為最優先，我會很感激。

This is really urgent, so please give it your full attention.
這相當緊急，所以請盡全力處理。

Unit
12

Vocabulary

allocate [ˈæləˌket] *v.* 分配；分派

Unit 13

描述和解決問題

n.p. doesn't work.
……無法運作。

例 I just got my laptop back from the repair shop, but now the touchpad doesn't work.

我剛從維修店拿回我的筆電,但現在觸控板無法運作。

The auto-dial on the fax machine doesn't work. And there's something wrong with the paper feed too.

傳真機上的自動撥號無法運作,送紙功能也有點問題。

n.p. won't work.
……無法運作。

例 I don't know why, but the printer just won't work.

我不知道爲什麼,但是印表機就是無法運作。

Don't bother bringing your cell phone to Japan. It won't work.

不必多此一舉地帶你的手機到日本,它無法運作。

n.p. doesn't work properly.
……無法正常運作。

例 I'm taking the printer back to the store because it doesn't work properly.
When I try to print, nothing happens; only this orange light blinks.

我要把印表機帶去店裡,因爲它無法正常運作。我要印時都沒反應,只有橘色的燈在閃。

For some reason the new PDAs they gave us just don't work properly.

不知爲何,他們給我們的新 PDA 就是無法正常運作。

Vocabulary

touchpad [ˋtʌtʃˌpæd] *n.*(筆電的)觸控板

n.p. won't work properly.
……無法正常運作。

例 If you don't keep the battery charged, it won't work properly.
如果你沒將電池充電，它就無法正常運作。

The copy machine won't work properly. The copies are too light and the paper keeps jamming.
影印機無法正常運作，影印的顏色太淡，而且一直卡紙。

n.p. doesn't work when(ever) + clause.
（每）當……就無法運作。

例 It doesn't work when I try to make double-sided copies.
當我要雙面列印時，它就無法運作。

The monitor doesn't work whenever I connect my PDA to the computer.
每次我把 PDA 連接到電腦時，螢幕就出問題。

n.p. doesn't work very well.
……運作得不是很好。

例 The new printer does work—it just doesn't work very well.
這台新印表機可以印，但運作得不是很好。

This computer is terrible. Even the keyboard doesn't work very well.
這台電腦糟透了，就連鍵盤也怪怪的。

I can't V.
我無法……

例 I have a built-in wireless card, but I still can't get online.
我有內建的無線網路卡，但我還是無法上網。

I can't figure out how to turn this thing off. Any ideas?
我不知道怎麼把這東西關掉，有任何主意嗎？

Unit
13

Vocabulary

built-in ['bɪlt'ɪn] *adj.* 內建的

figure out 理解；明白

I can't seem to V.
我好像不能……

例 I can't seem to connect my camera to the computer.
我好像無法把我的相機連接到電腦。

I can't seem to find the right driver for the webcam.
我好像無法找到網路攝影機的正確驅動程式。

n.p. doesn't seem to be Ving.
……好像無法……

例 The volume button doesn't seem to be working.
音量鈕好像無法運作。

The equipment doesn't seem to be responding to the commands.
這設備好像對指令沒有反應。

n.p. doesn't seem to V.
……似乎無法……

例 The screen doesn't seem to be bright enough.
螢幕似乎不夠亮。

The modem doesn't seem to work at all.
數據機好像根本不能運作。

n.p. seems to V.
……似乎……

例 It seems to jam whenever I try to print more than 50 pages at a time.
每當我一次要印超過五十頁好像都會卡紙。

The connection in the conference room seems to be a little weak.
會議室裡的訊號接收好像有點弱。

It seems that + clause.
……好像……

例 It seems that it's impossible to increase the production rate.
要提高生產率好像是不可能的。

302

It seems that we are having problems with the specs.

我們好像有規格方面的問題。

n.p. seems to have p.p./n.p.
……好像（已經）……

例 Our entire computer system seems to have crashed.

我們整個電腦系統好像已經故障。

The mail server seems to have a bad motherboard.

郵件伺服器的主機板好像有問題。

I seem to have p.p./n.p.
我好像（已經）……

例 I seem to have misunderstood your instructions.

我好像誤解了你的說明。

I seem to have a problem dealing with unhappy customers.

我在應付不滿意的客戶方面好像有問題。

I'm having trouble Ving.
我無法……

例 I'm having trouble understanding how we can increase quality and reduce expenses at the same time.

我無法理解我們要如何同時提高品質和降低費用。

I'm having trouble keeping the project on budget.

我無法將此專案控制在預算內。

I'm having trouble with n.p.
我對……有問題。

例 I'm having trouble with our steel supplier.

我和我們的鋼鐵供應商之間有問題。

I'm having trouble with the new online ordering system.

我不會使用新的線上訂購系統。

Unit
13

I'm having problems Ving.
我在……有問題。

例 I'm having problems communicating with our branch office in Bangkok.
我在和曼谷分公司的溝通上有問題。

I'm having problems arranging a meeting with the CEO. Maybe we should agree to see the vice president instead.
我無法安排和總裁的會面，也許我們應該改見副總裁。

I'm having problems with n.p.
我對……有問題。

例 I'm having problems with the new telephone system. I can't even get an outside line.
我使用新的電話系統時有問題，我甚至無法撥打外線。

I'm having problems with the client. They're threatening to find another supplier if we raise prices again.
我無法應付這客戶。如果我們再提高價格，他們威脅要找新的供應商。

n.p. be giving me problems.
……一直發生問題。

例 My new cell phone has been giving me problems. I'm probably just not used to it yet. 我的新手機一直發生問題，我可能是還不習慣吧。

They're giving me problems because I asked them to produce an extra 6,000 units by the end of the month.
他們一直找我麻煩，因為我要他們在月底前多生產六千件。

n.p. be giving me trouble.
……一直找我麻煩。

例 I'd like you to take over the Windermere account. Those guys have been giving me trouble for years.
我希望你接管溫德米爾帳戶，他們那些人幾年來一直找我麻煩。

I'm not completely sure, but I think it's the disk drive that's giving me trouble.
我不十分確定，但我想是磁碟機一直出問題。

Whenever I V, nothing happens.
每次我……，什麼都沒發生。

例 Whenever I press the delete button, nothing happens.
每次我按刪除鍵，什麼反應都沒有。

I don't like being mean, but whenever I ask politely, nothing happens.
我不想無禮，但每次我客氣地要求，沒人當一回事。

Whenever I V, it V.
每次我……，它……。

例 Whenever I plug my USB drive into my laptop, it crashes.
每次我把隨身碟插到筆電，它就當機。

Whenever I turn on my computer, it takes at least ten minutes to boot up.
每次我打開電腦，它至少得花十分鐘來開機。

 在這些 set-phrases 中，problems 和 trouble 的意義及用法是一樣的。

Unit
13

Vocabulary

crash [kræʃ] v.【電腦】當機   **boot up**【電腦】開機

What can I do?
我能怎麼做？

例 It's really important that I get in touch with her before 3:00. What can I do?
這很重要，我得在三點之前和她聯絡。我能怎麼做？

I really need Internet access to do my job. What can I do?
我真的需要上網才能工作，我能怎麼做？

How can I V?
我要如何……？

例 How can I solve the problem if both sides are not even willing to talk with each other? 我要如何解決這個問題，如果雙方不願意彼此溝通？

How do I V?
我要如何……？

例 How do I contact them?
我要如何聯絡他們？

How do I reset the defaults?
我要如何重設預設值？

How do you V?
你要如何……？

例 How do you get it to work?
你要如何讓它運轉？

How do you print to the color printer?
我要如何用彩色印表機列印？

Vocabulary

default [dɪˋfɔlt] *n.* 【電腦】預設值

306

How does it V?

它是如何……？

例 How does it run for so long on a single charge?

它如何充一次電就可運轉這麼久？

That's so cool! How does it work?

那很酷！它是如何運作？

How does n.p. work?

……如何運作？

例 How does the automatic sorting function work?

自動分類功能如何運作？

That's quite an impressive looking device you have there. How does it work?

你的那個設備看起來很炫，它要怎麼運作？

Is it possible to V?

有可能……嗎？

例 Is it possible to save my data before reinstalling the OS?

在重新啓動 OS 前有可能先儲存我的資料嗎？

I'd really like to set up a meeting as soon as possible. Is it possible to get together tonight after the conference?

我真的很想盡快安排會面，今晚在會議結束之後有可能碰個面嗎？

Any advice?

有任何建議嗎？

例 I'm having problems accessing my email from my new smartphone. Any advice?

我從我的新智慧型手機收取 email 有問題，有任何建議嗎？

I don't know what to do about the Braddock account. Any advice?

我不知道要怎麼處理 Braddock 的帳戶，有任何建議嗎？

Unit
13

Any suggestions?
有任何建議嗎？

例 Business has been really slow for the last two months. Any suggestions?
事業在過去兩個月發展得很慢，有任何建議嗎？

The battery isn't lasting as long as it used to. Any suggestions?
電池不像以往持久，有任何建議嗎？

Any advice you can give would be greatly appreciated.
任何你可以給的建議都很感激。

例 I know you're busy, but I'm really not sure how to handle the overstock problem. Any advice you can give would be greatly appreciated.
我知道你很忙，但我真的不確定要如何處理存貨過剩的問題。任何你可以給的建議都很感激。

I understand that you can't come into the office until next week, but any advice you can give would be greatly appreciated.
我理解你到下星期前都不能進辦公室，但任何你可以給的建議都很感激。

Please advise.
請提供建議。

例 I'm not sure what I should do with the returned merchandise. Please advise.
我不確定要如何處理退回的商品，請提供建議。

I need to know where to send the bill. Please advise.
我需要知道帳單要寄到何處，請提供建議。

Vocabulary

overstock [`ovə`stak] *n.* 存貨過剩　　**merchandise** [`mɝtʃən`daɪz] *n.* 商品；貨物

找出問題

n.p. may (not) be ...
……可能（沒有）……

例 Your computer may not be connected to the network. Check your connection.
你的電腦可能沒有連上網路，檢查一下你的連線狀態。

If the technician can't fix the problem on-site, the device may be faulty.
如果技術員不能當場修正問題，那麼儀器可能是有瑕疵的。

n.p. might (not) be ...
……可能（沒有）……

例 The disk drive might not be properly formatted.
磁碟機可能沒有適當地格式化。

The wiring might be loose. Have you opened up the service panel and checked that?
電線可能鬆了，你有打開伺服器面板檢查嗎？

... could be ...
……可能是……

例 There could be a problem with the way you've worded your email. Why don't you show me what you sent to him.
可能是你 email 的措辭有問題，你何不讓我看看你寄給他的東西。

The network could be overloaded. Why don't you wait and try again a little later in the day.
網路可能是超載，你何不等等、晚些時候再試看看。

Unit
13

Vocabulary

on-site [`an͵saɪt] *adv.* 當場　　　　**faulty** [`fɔltɪ] *adj.* 有缺陷的

n.p. may (not) have p.p.
……可能（沒有）……

例 They may not have included tax in the final price they quoted you.
他們最後給你的報價可能沒有含稅。

The prices may have gone down since June. Why don't we do another survey?
價格從六月以來可能已經下滑，我們何不另外做個調查？

n.p. might (not) have p.p.
……可能（沒有）……

例 The deliverymen might have installed it upside-down. 送貨員可能裝反了。

They might not have connected the printer to the right port.
他們可能沒有把印表機連接到正確的接口。

n.p. could have p.p.
……可能……

例 You could have pulled out the plug by accident.
你可能不小心把插頭拔掉了。

The CPU could have given out due to the heat.
CPU 可能因為過熱而停擺。

You may have forgotten to V.
你可能忘了……

例 You're not receiving the magazine because you may have forgotten to renew
your subscription. 你沒收到雜誌可能是因為你忘了續訂。

You may have forgotten to install the driver. 你可能忘了安裝驅動程式。

You might have forgotten to V.
你可能忘了……

例 You might have forgotten to include the figures from the previous month.
你可能忘了包含前一個月的數據。

I didn't get the file. You might have forgotten to attach it.
我沒有收到檔案，你可能忘了附加檔案。

Check n.p.
檢查⋯⋯

例 Check the connection.
檢查一下連線狀態。

Check the battery.
檢查一下電池。

Check that + clause.
檢查⋯⋯

例 Check that the correct drivers have all been installed.
檢查所有正確的驅動程式是否都已安裝。

If the total still looks wrong, check that you are using gross revenue, not net revenue.
如果總數看起來還是不正確,檢查一下你用的是毛利,而不是淨利。

Try Ving.
試著⋯⋯

例 Maybe you try unplugging the machine and letting it completely cool down before starting it again.
或許你試著把機器的插頭拔掉,讓它完全冷卻後再重新開機。

Try contacting them in the afternoon when they're not so busy.
試著在下午他們不那麼忙的時候聯絡他們。

Unit
13

Vocabulary

gross revenue 毛利

net revenue 淨利

unplug [ʌnˋplʌg] *v.* 拔掉插頭或塞子

Try again later.
稍後再試。

例 I'm sure they won't have time this morning. Why don't you try again later?
我確信他們今天早上不會有時間，你何不稍後再試？

Try again later. I might have some more information by then.
稍後再試，屆時我可能會有多一些的資訊。

Have you tried Ving?
你試過……嗎？

例 Have you tried restarting it? Well, have you tried kicking it?
你試過重新開機嗎？好吧，你試過踢它看看嗎？

Have you tried inviting him out for an informal lunch?
你試過私下邀請他吃午餐嗎？

You could try Ving.
你可以試著……。

例 You could try showing up at their office without an appointment. Just say
that you happened to be in the neighborhood and wanted to say hi.
你可以試著不預約就出現在他們的辦公室，就說你剛好在附近想來打個招呼。

You could try uninstalling the software and then reinstalling it.
你可以試著解除這個軟體的安裝，然後再重新安裝。

Did you remember to V?
你有記得……嗎？

例 Did you remember to make a reservation?
你有記得訂位嗎？

Did you remember to recharge the batteries?
你有記得幫電池充電嗎？

Vocabulary

show up 出現；露面　　　　　　　**happen to** 碰巧

What about Ving?

……如何？

例 What about transferring the files by FTP?
用 FTP 傳送這些檔案如何？

What about offering a slightly lower per-unit price instead of free shipping?
用提供低一點的單價取代免運費，如何？

How about Ving?

……如何？

例 How about doing the calculations again?
再重新計算一次如何？

How about asking the IT department what they think?
問問 IT 部門的想法，好嗎？

Why don't you/we V?

你們 / 我們何不……？

例 Why don't you call the help desk?
你何不打個電話給技術服務部門？

Why don't we upgrade the system so that we don't keep having this problem.
我們何不把系統升級，這樣就不會一直產生這個問題。

Make sure that + clause.

確保……

例 Make sure that they know we're flexible about everything except the price.
確保他們知道我們對所有事情都可有彈性，除了價格之外。

Make sure that all new employees know how to back their files up to the server.
確保所有新進員工都知道如何在伺服器上備份檔案。

Vocabulary

back up【電腦】備份

Let's V.
讓我們……

例 Your laptop is broken again? Well, let's have a look.
你的筆電又故障了嗎？好吧，讓我們看看。

Let's see if there's anything I can do to help.
讓我們看看有沒有我可以幫上忙的地方。

You must V.
你必須……

例 You must enter your password twice—once when you turn on the computer and again to access the database.
你必須輸入密碼兩次。一次是你打開電腦時，然後是要使用資料庫時。

You could V.
你可以……

例 You could have it repaired or return it and get a replacement.
你可以送修或是退回換貨。

You could ask Mr. Chen for his advice. I'm pretty sure he'd know how to handle this.
你可以問問陳先生的意見，我相信他知道如何處理這件事。

You can V.
你可以……

例 You can try reinstalling the program, but I doubt that it'll help.
你可以試試重新安裝這程式，但我不確定會不會有幫助。

You might V.
你或許……

例 You might want to upgrade to a newer model.
你或許想升級成較新的型號。

You might think about contacting the ombudsman to settle the dispute.
你或許想聯絡調查專員來解決爭端。

You should V.
你應該……

例 If you see smoke, you should unplug the computer immediately and contact the help desk right away.
如果你看到煙，你應該立刻拔掉電腦的插頭，然後馬上和技術服務部門聯絡。

You should call back on Monday when Julius is here. He should be able to help you.
你應該星期一朱力亞斯在這裡時再打電話過來，他應該可以幫助你。

You ought to V.
你應該……

例 You ought to get out of the office more. Why don't you spend a few days with me at the plant?
你應該多走出辦公室，你何不和我在工廠待幾天？

You ought to ask Mike for help. He's really good at that kind of thing.
你應該向麥可尋求協助，他真的很擅長那種事。

You need to V.
你應該……

例 You need to be very careful when negotiating with Mr. Liu. He knows all the tricks.
你和劉先生協商時必須很謹慎，他知道所有的策略。

After installing the software, you need to restart your computer.
安裝軟體之後，你必須把電腦重新開機。

Unit
13

... it should V ...
……它應該……

例 If you reset the defaults and restart it, it should work OK.
如果你重設預設值並重開機,它應該就會正常運作。

Unless you're really careless with it, it should run quite normally now.
除非你真的很草率地使用它,不然它現在應該運轉得十分正常。

... it could V ...
……它可以……

例 It could run faster if you added more memory.
如果你增加記憶體,它可以運轉得更快。

If you unplug it while it's running, it could damage the disk.
如果你在它運轉時拔掉插頭,那可能會損壞磁碟。

... it ought to V ...
……它應該……

例 They've already agreed to the price, so it ought to be easy to close the sale.
他們已經同意價格,所以這個交易應該很容易完成。

After it's been refurbished, it ought to run as good as new.
它在整修之後,應該可以運轉得像新的一樣。

... it will V ...
……它會……

例 With regular maintenance, it'll run fine for at least ten years.
有定期維修,它至少可以正常地運轉十年。

Stop worrying about your presentation. I'm sure it'll go smoothly.
別再擔心你的簡報,我相信它會很順利。

316

... you will V ...
……你會……

例 If you quit now, you'll never know if your plan would have worked.
如果你現在辭職，你永遠不會知道你的計畫是否可行。

When you hit Control + Alt + Y, you'll notice that a small dialog box appears.
如果你按「Control + Alt + Y」鍵，你會注意到有一個小對話框出現。

... you might V ...
……你或許……

例 By trying to save some money using that cheap content provider, you might spend more money in the end fixing the mistakes.
使用廉價的內容提供者來省錢，你最後或許得花更多錢來修正錯誤。

If you spend more time planning, you might find that you'll have fewer problems later.
如果你花多一點時間規劃，你或許會發現你之後會有較少的問題。

... it might V ...
……它可能……

例 It might be alright once you restart it.
一旦你重新開機，它可能就沒問題了。

It might work better if you move it into the shade.
如果你把它移到陰影下，它可能會運轉得好一點。

Unit 14

客戶管理與維繫

當客戶對服務表示滿意時

It [is always/has been] our pleasure to serve you.
為您服務〔永遠是／一直是〕我們的榮幸。

We are always available to serve you.
我們永遠在此為您提供服務。

Thank you for the opportunity to serve you.
感謝您讓我們有機會為您服務。

We appreciate having the opportunity to serve you.
我們很感激有機會為您服務。

希望未來繼續和對方保持生意往來時

We hope you'll let us serve you for many years to come.
希望您在未來也讓我們為您服務。

We will try our best to serve you in any way possible.
我們將以各種可能方式盡力為您服務。

We look forward to continuing to serve you.
我們很期待能繼續為您服務。

We look forward to the opportunity to serve you again.
我們期待能有機會再度為您服務。

客戶的問題已解決時

We hope we can serve you better.
希望我們能提供您更好的服務。

We've been pleased to serve you this past year.
我們很榮幸在過去的一年裡能為您服務。

We shall endeavor to serve you even better in the coming year.
我們會致力在新的一年裡提供您更好的服務。

We hope this change will serve your needs better.
我們希望此一改變能提供更符合您需求的服務。

We look forward to continuing to serve you from our new location.
我們希望在我們的新地址還能繼續為您服務。

「關於服務的 set-phrases」大部分的主詞是用 we 和 our。這是因為你是代表公司在寫 email。而且，使用可代表公司全體的 we 或 our，看起來更具有誠意。

Can you join me for n.p.?
你可以和我一起……嗎？

例 Can you join me for dinner on the 23rd?
你二十三日可以和我一起吃晚餐嗎？

Can you join me for lunch tomorrow?
你明天可以和我一起吃午餐嗎？

I invite you to join n.p.
我想邀請你加入……。

例 I invite you to join the yoga club. We do about 20 minutes of stretching during our lunch hour on Tuesdays and Thursdays.
我想邀請你加入瑜珈社團。我們在星期二和星期四的午休時間做約二十分鐘的伸展練習。

Maxutech will be celebrating our tenth year in business at the Riviera Club this Friday after work. I invite you to join the celebrations.
Maxutech 將在這個星期五下班後於 Riviera 俱樂部慶祝開業十週年。我想邀請你參加這個慶祝會。

I hope you can join me (for n.p.).
我希望你能一起來（……）。

例 I hope you can join me for a drink on Friday to celebrate my promotion.
我希望你星期五能一起來喝一杯，慶祝我獲得升遷。

I hope you will be able to join me (for n.p.).
我希望你能一起來（……）。

例 I hope you will be able to join me for my anniversary lunch this Friday.
我希望你這個星期五能夠一起來參加我的週年慶午宴。

I hope you will join me (for n.p.).
我希望你和我一起來（……）。

例 I hope you will join me for coffee after the presentation.
我希望你在簡報結束後跟我一起喝杯咖啡。

I'll be going for a coffee after the presentation. I hope you will join me.
我會在簡報結束後去喝杯咖啡，希望你一起來。

Please join me (for n.p.).
請和我一起（……）。

例 Please join me for a drink after work.
下班後請和我一起去喝一杯吧。

I'm going for a drink after work. Please join me.
我下班後要去喝一杯。請和我一起來吧。

Please join us (on n.p.).
請（在……）和我們一起來吧。

例 Please join us on Monday at 10:00 for the presentation of the sales awards.
請在星期一 10 點參加我們的業務人員獎勵大會。

The sales awards will be presented on Monday at 10:00. Please join us.
業務人員獎勵大會將於星期一 10 點舉行。請加入我們。

You're invited to join me (for n.p.).
邀請你和我一起（……）。

例 You're invited to join me for dinner on Thursday, May 23rd.
邀請你在五月二十三日星期四和我一起共進晚餐。

I'm hosting a dinner on Thursday, May 23rd. You're invited to join me.
我將於五月二十三日星期四主辦晚宴。邀請你一起參加。

Unit
14

在「關於邀請的 set-phrases」中主詞可用 I 和 me，也可用 we 和 us。
端看你是以公司或個人名義邀請，同時也得考量你和客戶的關係，如果
彼此熟稔，並已有長久的往來，則可使用 I 和 me 等較親近的語氣。

建立關係

表達祝賀

Congratulations on your promotion.
恭喜你升官了。

Congratulations on your wedding.
恭喜你結婚了。

I hope you will both be very happy!
希望你們兩位都能非常幸福！

Congratulations on the birth of your son/daughter.
恭喜你生了兒子／女兒。

Congratulations on your retirement.
恭喜你退休了。

Many happy returns!
生日快樂！

回應別人的好消息

That's great news!
這真是個好消息！

Well done!
做得好！

That's fantastic!
太棒了！

Good for you!
你真行！

回應別人的壞消息

Oh how awful!
噢，真糟糕！

Sorry to hear that.
很遺憾聽到這件事。

That's really terrible news.
那真是個不幸的消息。

You have my sympathies.
我很同情你。

表示同情

I'm sorry to hear of your resignation.
很遺憾聽到你辭職了。

I'm sorry to hear of your recent illness.
很遺憾聽到你最近生病了。

I hope you will recover soon.
希望你趕快好起來。

Get well soon!
早日康復！

I'm sure you will be back to normal soon.
我相信你很快就會復原的。

I'm sorry to hear of your recent bereavement.
很遺憾聽到你最近有親人過世。

We are all thinking of you here and send you our best wishes.
大家都在這裡為你集氣，並獻上最大的祝福。

Unit 15
關於附件

告知有附件

Please see the attachment.
請見附件。

n.p. is/are attached.
……已經附上。

例 The file you requested is attached.
你所要求的檔案已經附上。

The photos from the awards banquet are attached.
頒獎宴會的照片已經附上。

n.p. is/are included in the attachment.
……包含在附件中。

例 Contact information for everyone on the team is included in the attachment.
每一位組員的聯絡方式都包含在附件中。

Assembly instructions are included in the attachment.
組裝說明包含在附件中。

I am attaching n.p. for your consideration.
我附上……供你參考。

例 I am attaching our team's proposal for your consideration.
我附上我們團隊的提案供你參考。

I am attaching my resume for your consideration.
我附上我的履歷供你參考。

Vocabulary

assembly [ə`sɛmblɪ] *n.* 組裝

The n.p. is attached for your consideration.
附上……供你參考。

例 The proposal is attached for your consideration.
附上提案供你參考。

The first quarter sales report is attached for your consideration.
附上第一季銷售報告供你參考。

I have attached n.p.
我已附上……。

例 I have attached the report you requested.
我已附上你所要求的報告。

For your reference, I have attached some relevant figures.
供你參考，我已附上一些相關的數據。

Please find the attached n.p.
請見附件的……。

例 Please find the attached questionnaire. We thank you for your help in providing us with this information.
請見附件的問卷，感謝你協助提供我們這些資訊。

Please see the attached n.p.
請見附件的……。

例 For more detailed figures, please see the attached table.
更多的詳細數據，請見附件的表格。

Please see the attached proposal and let me know if you have questions.
請見附件的提案，讓我知道你是否有任何問題。

Unit
15

Vocabulary

relevant [ˈrɛləvənt] *adj.* 相關的   questionnaire [ˌkwɛstʃənˈɛr] *n.* 問卷

Please see the n.p. attached.
請見附件的……。

例 Please see the map of the conference venue attached.
請見附件的會場位置地圖。

Please refer to the attached n.p.
請參照附件的……。

例 Please refer to the attached schedule that includes information about the project timeline.
請參照附件的時程表，其中包含了專案時間規劃的資訊。

For more information about the product, please refer to the attached summary and supplementary materials.
更多關於此產品的資訊，請參照附件的摘要與補充資料。

Also attached is n.p.
還有，附件是……。

例 Also attached is a sample of my latest work.
還有，附件是我最新作品的樣本。

Also attached is a price list with both wholesale and suggested retail prices.
還有，附件是批發和建議的零售價目表。

Attached you will find n.p.
在附件中，你可以找到……。

例 Attached you will find my report on the situation in the Southern branch.
在附件中，你可以看到我針對南部分公司的情勢所作的報告。

Attached you will find contact details for our main distributors in your area.
在附件中，你可以找到你的區域內我們主要經銷商的聯絡資料。

Vocabulary

refer to 參考；查閱
wholesale [`hol,sel] *adj.* 批發的

supplementary [,sʌplə`mɛntəri] *adj.* 補充的
retail [`ritel] *adj.* 零售的

The following is/are n.p.
下列是⋯⋯

例 The following is my report.
下列是我的報告。

The following are my reasons for making this suggestion.
下列是我作出此一建議的理由。

... as follows: ...
⋯⋯如下：⋯⋯

例 Please find the figures as follows: 30 units at 250/unit, and 60 units at 230/unit.
請見數據如下：30 單位，則每單位 250 元計；60 單位，則每單位 230 元計。

Please find the proposal as follows: [insert proposal text]
請見提案如下：【插入提案內文】。

... the following: ...
⋯⋯下方：⋯⋯

例 In answer to your question, please see the following: X = 100 and Y = 300.
回答你的問題，請見下方：X= 100，Y= 300。

Please note the following: The venue has been changed to the Mount Vernon Hotel. 請注意下列事項：會場已變更至芒特佛南飯店。

... the following n.p.
⋯⋯下列⋯⋯

例 Please make sure that you carefully read the following documents regarding the case. 請務必仔細閱讀下列關於此案的文件。

She sent me the following article that you might find interesting.
她寄給我下列文章，你可能會覺得有趣。

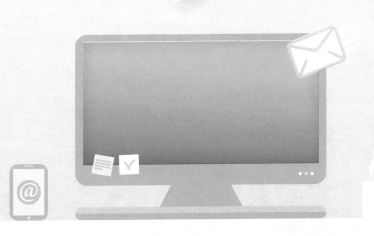

Unit 16
結語

頌候語

非常正式 （寫信給不認識的人、談及嚴肅主題時使用）

Yours faithfully,

Yours truly,

Yours sincerely,

專業 （寫信給認識的人，但保持些許的距離）

Best wishes,

Best regards,

Regards,

Sincerely,

專業又親切

Best,

Thanks,

Cheers,

1. 注意只有第一個字母大寫。
2. 注意逗號的使用。
3. Cheers 為英式用語。

Hope

I hope this is clear.
我希望這樣很清楚了。

I hope this helps.
我希望這有所幫助。

I hope this is OK.
我希望這樣沒問題。

I hope we can move this along quickly.
我希望我們能趕快開始進行。

I hope to hear from you soon.
我希望能很快得到你的消息。

I hope this is the beginning of a long and prosperous relationship.
我希望這是一段長期、成功關係的開始。

I hope to see you there.
我希望能在那裡見到你。

I hope you understand our position on this.
我希望你能理解我們對此事的立場。

Wish

I wish you the best of luck on the presentation.
我希望你簡報一切順利。

I wish we had more time to discuss this.
但願我們有更多時間討論此事。

I wish we had more time to prepare.
但願我們有更多時間準備。

I wish I could meet with you again before the product launch.
我希望在產品發售前能再與你碰面。

We wish you all a happy n.p.
我們祝福你們……快樂。

例 We wish you all a happy New Year.
我們祝福你們新年快樂。

We wish you all a happy holidays.
我們祝福你們假期愉快。

I wish you great success in n.p.
我祝你在……大獲成功。

例 I wish you great success in your new job.
我祝你在新的工作大獲成功。

I wish you great success in all your future endeavors.
我祝你所有未來的努力都能大獲成功。

Vocabulary

endeavor [ɪnˈdɛvə] *n.* 努力

包含 if 的結語

If you have any problems, ...
若你有任何問題，……。

例 If you have any problems, please don't hesitate to get in touch with me.
若你有任何問題，請儘管與我聯絡。

If you have any problems, just call me.
若你有任何問題，儘管打電話給我。

If you have any questions, ...
若你有任何問題，……。

例 If you have any questions, please feel free to contact me at any time.
若你有任何問題，請儘管隨時與我聯絡。

If you have any questions, just let me know.
若你有任何問題，儘管讓我知道。

If you have any questions, please do not hesitate to call either n.p. or n.p.
若你有任何問題，請儘管打電話給……或……。

例 If you have any questions, please do not hesitate to call either Jackie or me.
若你有任何問題，請儘管打電話給傑奇或是我。

If you have any questions, please do not hesitate to call either my mobile or my office number.
若你有任何問題，請儘管打我的手機或是辦公室電話。

Vocabulary

hesitate [ˈhɛzəˌtet] *v.* 猶豫；遲疑

Unit
16

You should contact n.p. if there are any problems.
若有任何問題，你可以聯絡……。

例 You should contact the helpline if there are any problems.
若有任何問題，你可以聯絡服務專線。

You should contact John in the marketing department if there are any problems.
若有任何問題，你可以和行銷部的約翰聯絡。

If you encounter any problems, contact me at n.p.
若你遇到任何問題，打到……和我聯絡。

例 If you encounter any problems, contact me at (02) 2314-2525.
若你遇到任何問題，請打到 (02) 2314-2525 和我聯絡。

If you encounter any problems, contact me at my hotel.
若你遇到任何問題，可以打到飯店和我聯絡。

Vocabulary

helpline [ˈhɛlplaɪn] *n.* 諮詢服務電話
encounter [ɪnˈkaʊntə] *v.* 遇到（困難、危險等）

Many thanks.
非常感謝。

Thanks for all your help.
謝謝你提供的一切協助。

Thank you very much (for your help).
非常謝謝你（的幫助）。

Thanks in advance (for your help/time).
先謝謝你（的協助／撥冗）了。

Thanks and sorry for any misunderstanding.
謝謝，並為造成的誤解致歉。

Many thanks for your understanding in this matter.
非常感謝你對此事的諒解。

Thank you for purchasing n.p.
感謝你購買……。

例 Thank you for purchasing the Speed-O-Flex exercise equipment.
感謝你購買 Speed-O-Flex 運動器材。

Thank you for purchasing our new product.
感謝你購買我們的新產品。

Thank you again for choosing n.p.
再次感謝你選擇……。

例 Thank you again for choosing our products and services.
再次感謝你選擇我們的產品與服務。

Thank you again for choosing XYZ Company.
再次感謝你選擇 XYZ 公司。

Unit
16

包含 appreciate 的結語

I appreciate all that you have done.
我感謝你所做的一切。

I appreciate you taking the time to help me with this.
我很感謝你撥空幫我處理此事。

I really appreciate the heads up you gave me the other day.
Thanks.
我非常感謝你日前給我的提醒。謝謝。

We appreciate it.
我們很感謝。

We appreciate your help.
我們很感謝你的協助。

We appreciate having you as a customer.
我們很高興能有你這樣的客戶。

We appreciate your business.
我們很感謝能與你做生意。

We appreciate your efforts.
我們很感謝你的努力。

We appreciate your interest in our company/products/
services.
我們很感謝你對我們的公司／產品／服務有興趣。

We appreciate your support.
我們很感謝你的支持。

Our apologies for the error.
我們為此錯誤致歉。

Our apologies for the delay.
我們為此延誤致歉。

I apologize for the inconvenience.
我為所造成的不便致歉。

I apologize for this regrettable error.
我為這個令人遺憾的錯誤致歉。

I apologize for this unfortunate error.
我為這個不幸的錯誤致歉。

We apologize for any inconvenience this may cause you.
我們為此可能造成你的任何不便致歉。

We apologize to you for any inconvenience we may have caused.
我們為我們可能造成的任何不便向你致歉。

Please accept our apologies for any inconvenience we may have caused you.
為了我們可能造成的不便，請接受我們的致歉。

附　錄

　　爲了實際工作上的方便，在此特別匯集了第一到第七章中使用率最高的語庫，讀者可依自身需求直接從中查找已分門別類的字串。

回信的開場白 set-phrases　P. 67

Thanking

* Thank you for purchasing n.p. ...
* Thank you for your message.
* Thanks for your reply.
* Thank you for your email about n.p. ...
* Thank you for your query about n.p. ...
* Thank you for sending me n.p. ...

Responding to Bad News

* We're sorry to learn (from X) of n.p. ...
* I'm sorry to learn (from X) about n.p. ...
* I am very sorry to learn (from X) that + n. clause ...
* I was sorry to hear about n.p. ...
* I was sorry to hear that + n. clause ...

Apologizing for Delay

* Apologies for the delay in getting back to you, but ...
* Sorry for the delay in Ving ...
* I apologize for the delay in getting back to you, but ...
* Apologies for the delay.
* Sorry for not getting back to you earlier, but ...
* Sorry for the delay in responding to ...

發信的開場白 set-phrases　P. 81

Announcing

* ... are included in the attachment.
* ... has been received.
* Also attached is n.p. ...
* Attached you will find n.p. ...
* For your information, ...
* I attach n.p. ...
* I have been informed that + n. clause ...
* Just to inform you of n.p. ...
* Just to inform you that + n. clause ...
* Just to let you know that + n. clause ...
* Please be informed that + n. clause ...
* Please be notified that + n. clause ...
* Please find the attached n.p. ...
* Please note that + n. clause ...
* Thank you for ... which I received today.
* This is just a quick note to let you know that + n. clause ...
* We have been asked to V ...
* We have just p.p. ...
* We have received n.p. ...

Reminding

* Have you remembered to V ...?
* Just a quick note to remind you to V ...
* Just to remind you that + n. clause ...
* Just to remind you to V ...
* Please remember to V ...

Making Confirmation

* I am pleased to confirm that + n. clause ...
* I can confirm that + n. clause ...

- I can confirm we have received ...
- Just a quick note to confirm n.p. ...
- Just a short note to confirm that + n. clause ...
- Just to confirm that + n. clause ...
- This is to confirm that I will be attending n.p. ...
- This is to confirm that + n. clause ...
- This is to confirm n.p. ...
- We hereby confirm that + n. clause ...
- We wish to confirm that + n. clause ...
- We wish to confirm the following: n. clause ...
- Your request has been successfully processed.

Requesting Confirmation

- Please confirm if + n. clause ...
- Please confirm our n.p. ...
- Please confirm receipt of n.p. ...
- Please confirm that + n. clause ...
- Please confirm whether + n. clause ...
- Please give me a call to confirm that + n. clause ...
- Would you please confirm that + n. clause ...?

說明背景的動詞 P. 116

achieve	have achieved	develop	have found
ask sb. to V		find	
become		finish	
change		send sth. to sb.	
complete		happen	
receive		improve	
speak to sb.		read	
write to sb.		look through	
decide to V			

346

說明原因的 chunks 和 words P. 118

as + n. clause	because + n. clause	because of n. p.
since + n. clause	so + n. clause	

說明期限的 chunks 和 words P. 128

Future Time	
◆ ... ASAP ...	◆ ... no earlier than ...
◆ ... as soon as possible ...	◆ ... in advance ...
◆ ... in a few days ...	◆ ... within the next few days ...
◆ ... next week ...	◆ ... at once ...
◆ ... by the end of this week ...	◆ ... later ...
◆ ... by the end of next quarter ...	◆ ... the day after tomorrow ...
◆ ... by the end of this quarter ...	◆ ... before COB today ...
◆ ... no later than ...	

Past Time	Past or Future Time
◆ ... a few days ago ...	◆ ... by the end of ...
◆ ... last week ...	◆ ... of the year ...
◆ ... yesterday ...	◆ ... in advance ...
◆ ... at the end of last quarter ...	◆ ... in time for ...
◆ ... at the end of last year ...	◆ ... soon ...
◆ ... the day before yesterday ...	◆ ... immediately ...
	◆ ... afterwards ...
	◆ ... by the 15th of ...

INFORMAL		
Making a Suggestion	**Rejecting a Suggestion**	**Accepting a Suggestion**
• Can you make n.p. ...? • Can you manage n.p. ...? • How about Ving ...? • How about n.p. ...? • How is ... for you? • Is ... OK for you? • Let's V ... • We could V ... • What about Ving ...? • What about n.p. ...? • Why don't we V ...? • Would ... be OK for you?	• ... is no good for me. • ... is not OK for me. • I can't make it then. • I can't manage it then. • My schedule is really tight. • My schedule is really tight + time.	• ... is good for me. • ... is OK for me. • I'll be free at ... • OK. • Sure.
FORMAL		
Making a Suggestion	**Rejecting a Suggestion**	**Accepting a Suggestion**
• At this point I'd like to propose that + n. clause ... • At this point I'd like to propose n.p. ... • I'd like to suggest n.p. ... • I'd like to suggest that + n. clause ... • What time would suit you? • When would be convenient for you?	• I'm afraid I have to V ... • I'm afraid I'm going to be Ving ... • I'm afraid I'm going to be ... • I'm afraid I'm tied up then. • I'm afraid that's not possible.	• That would be fine. • That's fine. • That's good for me. • Yes, I can make it. • Yes, I can manage that.

確認關於會面的事項 P. 148

Describing	Changing
◦ ... be going to V ...	◦ I need to change/cancel/postpone n.p. ...
◦ ... be Ving ...	◦ I'm afraid I have to V ...
◦ I'll be arriving ...	◦ I'm afraid we're going to have to cancel n.p. ...
◦ I'll be leaving ...	◦ I'm afraid we're going to have to postpone n.p ...
◦ I'll be Ving ...	
◦ I'll have to V ...	◦ I'm going to have to V ...
◦ I've got a meeting ...	◦ Would it be OK if we postponed it till ...?
◦ I've got an appointment ...	

包含 will 的常見 set-phrases P. 157

Making Offers		Making Requests
◦ I'll get it.	◦ I'll do it.	◦ Will you do it?
◦ I'll V ...	◦ Shall I do it?	◦ Will you help me with this?
◦ I'll see what I can do.		

Making Promises		Other
◦ I'll give you a ring.	◦ I'll V ...	◦ It'll be alright.
◦ I'll drop you a line.	◦ I'll be in touch.	◦ It'll take time.
◦ I'll send it tomorrow.		◦ We'll see.
◦ I'll be free at ...		◦ I think I'll probably V ...
◦ I'll get back to you as soon as I can.		◦ I think X will V ...
◦ I'll be back on Thursday.		◦ I think it'll V ...
◦ She'll be back tomorrow.		◦ I'll be late.
◦ I'll see what I can do.		
◦ I'll see you then.		

Describing Problems	Asking for Help
• I can't V ...	• What can I do?
• I seem to have p.p./n.p. ...	• What should I do?
• I'm having problems Ving ...	• How do you ...?
• I'm having problems with n.p. ...	• How does it ...?
• I'm having trouble Ving ...	• How does X work?
• I'm having trouble with n.p. ...	• How can I ...?
• ... be giving me problems.	• How do I ...?
• ... be giving me trouble.	• Any advice you can give would be greatly appreciated.
• ... doesn't work.	
• ... doesn't work properly.	• Any advice?
• ... doesn't work very well.	• Any suggestions?
• ... doesn't work when(ever) + n. clause ...	• Is it possible to V ...?
• ... won't work.	• Please help.
• ... won't work properly.	• Please advise.
• ... won't work when(ever) + n. clause ...	
• I can't seem to V ...	
• ... doesn't seem to V ...	
• ... doesn't seem to be Ving ...	
• ... seems to V ...	
• It seems that + n. clause ...	
• ... seems to have p.p./n.p. ...	
• Whenever I V ..., nothing happens.	
• Whenever I V ..., it V ...	

解決問題的 set-phrases 和 chunks (1) P. 175

Finding the Cause of the Problem	
• ... check n.p. ...	• ... may (not) have p.p. ...
• ... check that + n. clause ...	• ... might (not) have p.p. ...
• You might have forgotten to V ...	• ... could have p.p. ...
• ... may (not) be ...	• You may have forgotten to V ...
• ... might (not) be ...	• Have you tried Ving ...?
• ... could be ...	• Have you remembered to V ...?

Suggesting a Solution	Recommending Action
• You could try Ving ...	• ... check n.p. ...
• What about Ving ...?	• ... check that + n. clause ...
• How about Ving ...?	• ... try Ving ...
• Why don't you/we V ...?	• ... try again later.
	• Make sure that + n. clause ...
	• Let's V ...
	• See if you can V ...

解決問題的 set-phrases (2) P. 179

Options	Actions	Consequences
... you could V ...	... you should V ...	... it should V ...
... you can V ...	... you must V ...	... it could V ...
... you might V ...	... you ought to V ...	... it ought to V ...
	... V ...	... you will V ...
	... you need to V ...	... it will V ...
		... you might V ...
		... it might V ...

Countable	Uncountable	Both
dollar	money	work
suggestion	help	paper
person	machinery	time
machine	merchandise	
fact	information	
job	research	
man	advice	
equipment	input	
people	news	
newspaper	feedback	
	staff	

- If you have any problems, ...
- If you have any questions regarding this, please let me know.
- If you require further assistance, please feel free to contact me at any time.
- If you have any further questions or concerns, please feel free to contact us at any time.
- If you have any questions, please do not hesitate to call either ... or ...
- If you have any questions about this, please do not hesitate to contact me.
- You should contact ... if there are any problems.
- If you face any problems, contact me at ...
- If you have any problems, don't hesitate to contact ...

其他種類的結尾 set-phrases P. 224

- I look forward to meeting you again soon.
- Have a good day.
- I look forward to your reply.
- Have a good weekend.
- I look forward to working with you on this project.
- Talk to you soon.
- I look forward to hearing from you soon.
- Feel free to contact me at any time.

包含 appreciate 的結尾 set-phrases P. 231

Thanking for Help

- I really appreciated your help the other day.
- We are always very appreciative of your efforts.
- We very much appreciate your help with this.
- Please allow me to express my appreciation for your help.

Asking for Help

- I would appreciate your help with this.
- Your help with this would be much appreciated.
- We would very much appreciate it if you could help us.

Rejecting a Request for Help

- I appreciate your concerns, but I'm afraid I am not able to help you.

延伸閱讀建議 P. 58

send/request/convey/clarify information
寄 / 要求 / 傳達 / 釐清訊息

例 Why didn't you send the information to me as soon as you received it?
你為什麼沒有在收到訊息時就馬上寄給我？

The information you requested can be found in our annual report, which is available for download on our website.
你要求的資訊在我們網站可供下載的年報中可以找到。

Please don't convey any information about our production problems to the customers.
請不要傳遞關於我們生產問題的訊息給客戶。

You should include some graphs in your presentation to help clarify the information.
你應該在簡報中包含一些圖表以利釐清訊息。

confirm/make an arrangement
確認 / 安排

例 We agreed to share a hotel room at the conference, but I haven't confirmed the arrangements with her yet.
我們同意在會議期間共用一個旅館房間，但我尚未和她確認這些安排。

I've made arrangements for you to be picked up from the airport.
我已作了安排，有人會到機場接你。

Vocabulary

convey [kən've] v. 傳達；傳遞

graph [græf] n. 圖表

make/offer/reject/respond to/accept a suggestion
提出 / 提出 / 推翻 / 回應 / 接受建議

例 I'd like to make a few suggestions about the way our products are marketed in Europe.
關於我們的產品行銷到歐洲的方式，我想提出一些建議。

I'd like to offer a suggestion here. Why don't we include a picture of the product?
在此我想提出一個建議。我們何不加入一張產品的照片？

I wanted to offer a bigger discount for bulk purchases, but they rejected my suggestion.
對於大宗採買，我想提供更大的折扣，但他們拒絕我的建議。

If she doesn't respond to your suggestion within a week, you can pretty much take that as a no.
如果她在一星期內沒有對你的建議作出回應，你大概可以把它當成是拒絕。

If we accept your suggestion, how do you propose we pay for it?
如果我們接受你的提議，你建議我們如何支付費用？

describe/outline/initiate a procedure
描述 / 概述 / 啓用程序

例 Could you describe the procedures you use to identify defective products?
你能描述你用來找出瑕疵產品的程序嗎？

Let me just outline our procedures for maintaining the customer database.
讓我概述一下我們維護客戶資料庫的程序。

We've decided to initiate some new quality control procedures.
我們已經決定啓用一些新的品管程序。

Vocabulary

bulk purchase 大量購買
outline [ˈautˌlaɪn] v. 概述

reject [rɪˈdʒɛkt] v. 拒絕；抵制
initiate [ɪˈnɪʃɪˌet] v. 開始；發起

offer/extend/decline (= turn down = refuse)/respond to/accept an invitation
提出 / 提出 / 拒絕 / 回應 / 接受邀請

例 I wonder why they didn't offer you an invitation to join the club. I got one.
我納悶他們爲何不邀請你參加俱樂部。我受到邀請了。

I'd like to extend an invitation to all of our regional managers to visit us here in Taipei on the first Wednesday of every month.
我想要邀請所有的區域經理,在每個月的第一個星期三來台北拜訪我們。

If the CEO wants you to go with her, I really don't think you should decline her invitation.
如果總裁要你和她一起去,我眞的覺得你不應該拒絕她的邀請。

I'm not sure if I can make it so I haven't responded to the invitation yet.
我不確定我是否能去,所以我尚未回覆這個邀請。

If you can't accept her invitation, you should at least write her a short note to explain why.
如果你不能接受她的邀請,你應該至少寫封簡短的信解釋爲什麼。

outline/put forward/implement/announce a plan
略述 / 提出 / 實施 / 公布計畫

例 I just briefly outlined my plan at the meeting. The details are in the report.
我在會議中只概述我的計畫,細節則在報告中。

Good ideas are a dime a dozen. If you want funding, you'll have to put forward a comprehensive plan.
好的意見並不值錢。如果你想取得資金,你必須提出一個全面的計畫。

Announcing a plan is a lot easier than implementing it.
公布一個計畫比實行一個計畫容易許多。

Vocabulary

decline [dɪˋklaɪn] *v.* 婉拒;謝絕
funding [ˋfʌndɪŋ] *n.* 資金;基金

a dime a dozen【口語】 多得不稀罕的;不值錢的
comprehensive [͵kɑmprɪˋhɛnsɪv] *adj.* 綜合的;全面的

make/respond to/deal with a request
提出 / 回應 / 處理要求

例 You have to make a written request in order to be reimbursed for travel expenses.
你必須提出書面要求來申請你的差旅費用。

It's company policy to respond to all customer service requests within twenty-four hours. 公司的政策是要在 24 小時內回覆所有的客服要求。

If you don't know how to deal with an unusual request from a customer, just pass it on to the team leader.
如果你不知道如何處理客戶的特殊要求，就把它轉給組長。

make/offer/reject/respond to/accept a proposal
提出 / 提出 / 拒絕 / 回應 / 接受提案

例 If you're not satisfied with the wording of the contract, please make a proposal describing the specific changes you'd like to see.
如果你對合約的措辭不滿意，請提出提案具體說明你想修正的地方。

I'd like to offer a proposal that I think both sides can accept.
我想要提出一個我認為雙方都可接受的提案。

If I hadn't asked for such a large budget, they probably wouldn't have rejected my proposal.
如果我不要求這麼大筆的預算，他們或許不會拒絕我的提案。

Give them some time to respond to your proposal. They probably just need to think about it some more.
給他們一些時間來對你的提案作出回應。他們或許只是需要再多想想。

I'm sure they'll accept your proposal. It doesn't cost anything and could end up saving the company a lot of money.
我相信他們會接受你的提案。它不需花費任何成本，而且最後可能會替公司省下一大筆錢。

Vocabulary

reimburse [ˌriɪmˈbɝs] *v.* 償還；退款

specific [spɪˈsɪfɪk] *adj.* 明確的；具體的

describe a process
描述過程

例 The attached file describes the reimbursement process. Please read it before filling out the request form.

附件描述了退款的程序，填寫申請表格前請先讀一下。

give an instruction
給指示

例 I gave you clear instructions, which you did not follow. You're lucky the whole factory didn't burn down.

我給了你清楚的指示，但你沒有遵守。你很幸運整間工廠沒有付之一炬。

raise/ask/answer/respond to a question
提出 / 問 / 回答 / 回應問題

例 Your attitude lately has raised a few questions about your loyalty to the company.

你最近的態度反應出你對公司忠誠度的一些問題。

Can I ask a question about your long-term goals? Where do you see yourself in five years?

我可以問個關於你長期目標的問題嗎？你覺得你五年後會是什麼樣子？

If you can't answer a customer's question, please don't just make something up. Ask a supervisor instead.

如果你不知道如何回答顧客的問題，請不要憑空捏造。問一下主管。

Any successful sales person knows how to respond to a question without really answering it.

任何成功的業務員都知道如何不正面地回應問題。

provide background
提供背景

例 In this section of the report, I'd like to provide some background on our target demographic.

在報告的這一部分，我想提供我們目標族群的一些背景資料。

make/reject/respond to/accept an offer
提出 / 拒絕 / 回應 / 接受提議

例 They made an offer but it wasn't very reasonable. Anyway, we've decided not to sell. 他們提出一個報價，但並不是非常合理。總之，我們決定不賣。

They rejected our first offer, so we'll raise it a little bit and try again. 他們拒絕我們的第一個報價，所以我們提高一些並再試一次。

They haven't responded to our offer yet. They're probably waiting to see how anxious we are. 他們尚未回應我們的提議，他們可能在等著看我們會有多緊張。

We've decided to accept your offer. We'll be in touch to discuss details with you. 我們決定接受你的提議，我們會與你聯繫討論細節部分。

make/deal with (= handle)/settle a complaint
提出 / 處理 / 解決抱怨

例 I'd like to make a complaint about the service I received from one of your salespeople. 我想對你們一位售貨員的服務提出抱怨。

I'll be dealing with your complaint from now on. 從現在起我會處理您的投訴。

It took them ages to settle my complaint. I won't be doing business there again. 他們花了好長的時間來解決我的投訴，我不會再到那裡做生意了。

give a reason
給理由

例 If a customer wants to return a product, please don't ask them to give a reason. Just smile and return their money. 如果消費者想要退貨，請別要求他們給理由。只要微笑並退錢給他們就好了。

request/take action
要求 / 採取行動

例 The client is requesting action on this. What should I tell her? 客戶要求我們對此採取行動。我應該怎麼告訴她？

We can't ignore the over-65 market any longer. If we don't take action immediately, it'll be too late.

我們不能再忽略 65 歲以上的市場。如果我們不立刻採取行動，就會太遲了。

give/set a due date (= deadline)
給 / 設定截止日

例 The due date I gave you earlier doesn't appear to be realistic. Let's try to have it finished by next Thursday instead.

我早先給你的截止日好像不切實際。我們改試著在下星期四前完成吧。

Why is the schedule such a mess? Because management sets the deadlines without consulting the people actually doing the work.

為什麼時程表一團糟？因為管理者沒有和實際執行工作的那些人商量就設定截止日。

give/request/demand an explanation
給 / 要求 / 要求解釋

例 Even if you don't know what went wrong, you still have to give the customer some kind of explanation.

就算你不知道出了什麼錯，你還是得給客戶一個解釋。

I'm writing to request a written explanation of what happened to my order.

我寫信來要求針對我的訂單狀況提出一個書面解釋。

This is the third time that the shipment has been delayed. I demand an explanation!

這已經是第三次送貨延遲了，我要求一個解釋！

make (a) reference to
參考

例 In your report you make a reference to new government regulations. Can you talk a little bit more about how they could affect us?

你在報告中參考了新的政府規定。你可以多談一下它們會如何影響我們嗎？

change/cancel/place an order
更改 / 取消 / 下訂單

例 I'm writing to change my order. Please make it six cases, rather than four.
我寫信來更改訂單。請改成 6 箱，而不是 4 箱。

I'm writing to cancel my order. We actually have enough in stock at the moment.
我寫信來取消我的訂單，事實上我們現在有足夠的庫存。

I'd like to place an order for 20 units, please.
我想要下 20 件的訂單，麻煩了。

國家圖書館出版品預行編目資料

商務英文 Email 速成語庫書 / Quentin Brand 著;戴至中
譯 . -- 初版 . -- 臺北市;貝塔,2015.06
　　面:　　公分
　　ISBN 978-957-729-988-8（平裝）
　　1. 商業書信　2. 商業英文　3. 商業應用文　4. 電子郵件

493.6　　　　　　　　　　　　　　　　　104006262

商務英文 Email 速成語庫書

作　　者 / Quentin Brand
譯　　者 / 戴至中
執行編輯 / 游玉旻

出　　版 / 波斯納出版有限公司
地　　址 / 100 台北市館前路 26 號 6 樓
電　　話 / (02) 2314-2525
傳　　真 / (02) 2312-3535
客服專線 / (02) 2314-3535
客服信箱 / btservice@betamedia.com.tw
郵撥帳號 / 19493777
郵撥戶名 / 波斯納出版有限公司

總 經 銷 / 時報文化出版企業股份有限公司
地　　址 / 桃園市龜山區萬壽路二段 351 號
電　　話 / (02) 2306-6842

出版日期 / 2021 年 1 月初版四刷
定　　價 / 420 元
I S B N / 978-957-729-988-8

貝塔網址:www.betamedia.com.tw

喚醒你的英文語感！

Get a Feel for English !

喚醒你的英文語感！

Get a Feel for English !